UNDERGROUND

"[An] absorbing narrati

Book Review

"Rudd conveys well the festival-like joy ... time campus uprisings of the late sixties. . . . Rudd's historical judgments are, to use a phrase from the era, right on." —*Los Angeles Times*

"A gem. . . . Even those who condemn Rudd's work in history can be grateful for Rudd's work of history." —*Washington Post*

"Rudd is reflective and truthful. . . . He shatters the romantic myth of the Weather Underground." —*New York Post*

"A trailblazer. . . . Rudd's essential contribution is his self-portrait as a youth who persuaded others to wreck rather than create—and his snapshots of like-minded contemporaries." —*Wall Street Journal*

"This important memoir gives an eyewitness account into some of the most divisive years in American history. Rudd provides valuable insight into those years." —*Albuquerque Journal*

"The value of *Underground* is not to be measured by the depth of its self-criticism. Rather, it is worth reading as a travel guide into hell, a story-lesson, and a warning about the risks of ideology inherent in all militant activism." —*New York Observer*

"In this account Rudd carefully sorts 'what was right from what was wrong' with honesty, regret, and hard-won wisdom."

—*Publishers Weekly*

"A provocative memoir. . . . If you thought the right wing was in a lather over Bill Ayers, wait until its talking heads get hold of this unapologetic book, which deserves to be read and discussed."

—*Kirkus Reviews*

"Rudd's account stands with the best of these works, deftly describing the banality of life inside the movement. Rudd retains his idealism and rebellious spirit but sees positive action and community building as the true path forward. This one's a good read."

—*Library Journal*

"Rudd is not on some flashback trip through 'Strawberry Fields Forever.' Nor is his often-riveting new memoir an exercise in nostalgia, apologia, or retread rhetoric. It is a stark look back in candor. Rudd is searingly frank and self-critical, especially about the Weathermen's disastrous ventures into violent Days of Rage in Chicago and bombings galore later."

—*Daily Beast*

"Refreshingly honest, Rudd writes about his road to radical. . . . Rudd, unlike Bill Ayers, is clear about what he regrets, and what he would have done differently."

—Annette John-Hall, *Philadelphia Inquirer*

UNDERGROUND

HARPER

NEW YORK • LONDON • TORONTO • SYDNEY

Mark Rudd

UNDERGROUND

My Life with SDS and the Weathermen

HARPER

Grateful acknowledgment is made for permission to reprint the following: insert page 2: Sundial Rally, IDA 6, *Richard Howard*; insert page 3: top: Dean Coleman and Mark Rudd, *Steve Ditlea;* middle: "JOIN US" leaflet, *Columbia University Archives, 1968*; bottom: David Gilbert, *Richard Howard*; insert page 4: Juan Gonzalez and Mark Rudd, *David Finck*; insert page 5: top: Meeting in Fayerweather, *Gerald Adler*; bottom: police line, *Richard Howard*; insert page 6: top: Amsterdam Gate rally, *Nicholas Mirra*; bottom: SAS march, *Richard Howard*; insert page 7: top: "Megaphone Mark," *Columbia University Archives, 1968*; bottom: Ted Gold, *Richard Howard*; insert page 8: top: Bertha and Mark Rudd, Edward Hausner/*New York Times*; bottom: "Confrontation at Columbia" cover, courtesy of *Newsweek*; insert page 9: top: Kansas University speech, *Kansas University Archives;* bottom: Bernardine Dohrn, *David Fenton*; insert page 10: top: Weatherman line, *David Fenton;* bottom: Days of Rage arrests, *David Fenton*; insert page 11: Town house: *Fred McDarrah*; insert page 12: bottom: "The Soiling of Old Glory," Pulitzer Prize, 1977, on *Osawotamie* cover, *Stanley J. Forman*; insert page 14: 100 Centre St., *Bettman/Corbis*; insert page 15: bottom: Nicaragua construction brigade, *Allen Cooper*; and insert page 16: Drew University talk, *Thomas Good/Next Left Notes.* All other photographs from the author's family collection.

A hardcover edition of this book was published in 2009 by William Morrow, an imprint of HarperCollins Publishers.

FIRST HARPER PAPERBACK PUBLISHED 2010.

Photographic illustration © Bram Janssens|Dreamstime.com
Designed by Kate Nichols

Library of Congress Cataloging-in-Publication Data has been applied for.

ISBN 978-0-06-147276-3

10 11 12 13 14 OV/RRD 10 9 8 7 6 5 4 3 2 1

Contents

Preface

For twenty-five years I'd avoided talking about my past. During that time I had made an entirely new life in Albuquerque, New Mexico, as a teacher, father of two, intermittent husband, and perennial community activist. But in a few short months, two seemingly unrelated events came together to make me change my mind and begin speaking in public about my role in Students for a Democratic Society (SDS) and the Weather Underground—things that happened to me when I was a kid.

First, in March 2003, the United States attacked Iraq, beginning a bloody, long, and futile war of conquest. What I saw, despite some significant differences, was Vietnam all over again. As a reflex I joined the antiwar movement with millions of others, just as I had done thirty-eight years before, when I was eighteen years old.

From 1965 to 1968, the years of the big escalation of the Vietnam War and the maturation of the civil-rights movement,

I was a member of SDS at Columbia University in New York City, one among many hundreds who made as much noise and trouble as possible to protest the university's pro-war and racist policies. The organizing was good and the time was right, so the campus blew up in April 1968 with the largest student protest up to that point. Having been recently elected chairman of the Columbia chapter of SDS, I was identified by the press as the strike's top leader; the impudent young twenty-year-old with the megaphone. The cartoonist Garry Trudeau even created a *Doonesbury* character modeled after me, Megaphone Mark, a true icon of the sixties.

As both the Vietnam War and the antiwar movement grew red hot, I went over the cliff with a tiny fragment of the much larger SDS. We thought we were living a line from the Rolling Stones song "Street Fighting Man": "Think the time is right for Palace Revolution / But where I live the game to play is Compromise Solution." My friends and I formed an underground revolutionary guerrilla band called Weatherman which had as its goal the violent overthrow of the United States government. Confirmed idealists, we wanted to end the underlying system that produced war and racism. It didn't work. From 1970 to 1977, I was a federal fugitive, living the whole time underground inside this country.

Just a few months after the Iraq War began, the documentary movie *The Weather Underground* was released. The project of two young men then in their thirties, Sam Green and Bill Siegel, the movie had been more than five years in the making. I am featured both as a contemporary talking head and also in archival footage as a twenty-year-old revolutionary. Nominated for an Academy Award and broadcast nationally on PBS, the movie has been a great success with audiences and critics; it remains in circulation as a DVD and continues to be shown frequently in college and high-school classes, stimulating much comment and many questions.

The closing images of the movie show me as a befuddled, gray-haired, overweight, middle-aged guy observing that thirty years later I still don't know what to do with my knowledge of who we are in the world; then the film cuts to aerial shots of carpet bombing in Vietnam and, finally, to a close-up of a skinny twenty-year-old kid, the same guy, with the same grief-stricken look on my face. This ending hits audiences like a blade going right to the existential gut of our problem.

In the years since 2003, I've spoken and answered questions at scores of colleges, high schools, community centers, and theaters about why my friends and I opted for violent revolution, how I've changed my thinking and how I haven't, and, most of all, about the parallels between then and now. Young audiences are hungry to know this history, sensing its relevance to today. They seem genuinely amazed to learn that once there was a group of young white kids from privileged backgrounds who risked everything for our antiwar, antiracist, and revolutionary beliefs, to act "in solidarity with the people of the world."

Sometimes passion doesn't rule the day. In *The Weather Underground*, I say I haven't wanted to talk about my past because of my "guilt and shame." I never get to explain what I'm guilty and ashamed of, but it's implied: Much of what the Weathermen did had the opposite effect of what we intended. We deorganized SDS while we claimed we were making it stronger; we isolated ourselves from our friends and allies as we helped split the larger antiwar movement around the issue of violence. In general, we played into the hands of the FBI—our sworn enemies. We might as well have been on their payroll. As if all this weren't enough, three of my friends died in an accidental explosion while assembling bombs. This is not a heroic story; if anything, it's antiheroic.

Having made such disastrous mistakes on such a big level—even granted that I was twenty years old at the time—I spent decades doubting my judgment. It took me a long time to sort out what was right from what was wrong in my own history.

But in conversations with young people since 2003, I've found that Weatherman's failures are less important to them than the simple astonishing fact that we existed. As a result of this ongoing dialogue,

I've shifted my opinion about my own past, and in doing so I've rediscovered a voice that I bottled up for two and a half decades—longer than most of the people I was speaking to have been alive. I've also reclaimed what I can be proud of: Along with millions of other people, I was part of a movement of history—that's what a "movement" is, after all, a shift of history caused by millions—that helped end the war in Vietnam. Combined with the civil-rights movement, the period was American democracy's finest hour. Historical movements aren't made of heroes, just ordinary people trying to do right. The movements of the sixties succeeded in transforming laws and practices concerning the position of black and other minority people in this country, and they helped stop a major war of aggression by our own government

That sucessful mass movements happened in my lifetime tells me they can happen again. The election of Barack Obama has liberated young people's political imagination and energy. I hope my story helps them figure out what they can do to build a more just and peaceful world. At the very least, it might show some serious pitfalls to avoid.

PART I

Columbia

(1965–1968)

1

A Good German

My mother tells this story about dropping me off at the dorm at West 114th Street and Broadway on the first day of Freshman Week. She and my father and I were unloading the car when a kid came up to me and handed me a blue and white beanie, the official headgear of the Columbia College freshman. He said, "We have to wear it all week."

I replied, "That's stupid, I'm not gonna wear that thing!"

Bertha, my mother, looked at Jake, my father, and said quietly, "We're in for trouble."

Actually, my mother had it wrong about my refusal to wear the beanie. It wasn't instinctive rebelliousness; I just didn't want to act like a kid. I had dreamed of this moment throughout my childhood in suburban New Jersey, longing to go off to college and finally not have to pretend to be a child. I'd always felt like a misfit with other kids. Play didn't interest me: I liked to read—history, biography, science, novels—and to work, to chop firewood, to build things. At last I had escaped the loneliness and shame of childhood, and I didn't want this, my coming-of-age moment—my true bar mitzvah, the day I was supposed to become a man—ruined by anything so juvenile as a stupid beanie.

I spent those first months at Columbia roaming the campus and glorying in the great classical brick-and-limestone-faced buildings, the columned libraries, even the herringbone-patterned brick sidewalks. I was awestruck to be part of this mighty international university.

Columbia was built upon one of the highest points in Manhattan, first called Harlem Heights and later Morningside Heights. An early battle of the Revolutionary War, in which the Americans finally proved themselves, had taken place here. Morningside Heights looked out over Harlem, a vast valley of apartment buildings, mostly walk-up tenements, extending miles to the east and north, at the time the largest black ghetto in the United States. Columbia University was the crown set atop the Heights. At the loftiest point on the campus, the central visual focus, loomed the monumental Low Library, the seat of the university's administration, immodestly modeled after the Roman Pantheon, its enormous columns and huge rotunda the symbol of imperial power. All this was *mine* now.

A few times a week, I would go to class wearing a blue blazer, the official uniform of Columbia College men. Deans held afternoon socials with the students, during which we sat around drinking sherry from crystal goblets. Yes, that was me, a Jewish *pisher* from the New Jersey suburbs, in a leather armchair, sipping sherry and chatting with a WASP assistant dean about Plato in an oak-paneled lounge like no other room I'd ever been in. Of course I'd never tasted sherry either.

But something gradually began to feel wrong. I'd be sitting in my freshman English class, learning to analyze nineteenth-century British poetry, and suddenly I would be overcome by a wave of despair.

Confused questions would pop into my mind: Why am I here, scrunched into a tiny wooden desk in this overheated classroom, pretending to be interested in poetry? Who are these boys sitting next to me in their blue blazers, regimental ties, and pressed slacks? And I also wondered like many an eighteen-year-old guy, why can't I sleep with every girl I meet?

I wandered into the Columbia College Counseling Service, looking for help. When the counselor asked what was wrong, I told him I was having trouble studying and paying attention in class. "Ah, Freshman Identity Crisis," he said, probably having heard the same story six times that day. I hadn't known there was a name for what I was going through. He asked about my sex life. I said I was depressed about Liliana, my high-school girlfriend who was now at Sarah Lawrence College. I loved her and wanted to stay with her but also wanted to sleep with other women.

He was paying attention now. "So you are having sexual problems," the counselor said. "You would most certainly benefit from analysis."

I thought about this. I had read Freud in high school, and his method had intrigued me: interpretation of dreams, the tripartite personality structure, the whole schmear about the unconscious. Psychoanalysis was the intellectual, even bohemian thing to do in New York City in the forties, fifties, and sixties. It was European. On the other hand, no one in my family had ever been to a shrink, and I wasn't sure I wanted to be the first.

The counselor was way ahead of me. "Are you receiving financial aid to attend Columbia?" he asked.

"No, my parents are paying the whole tuition."

His eyes lit up. "I'd like to refer you to Dr. Robert Liebert, a psychiatrist on our staff. He also sees patients privately."

I spent the Christmas vacation moping around my empty dorm, alone, depressed, not wanting to go home to Maplewood for more than an obligatory half-day visit. When I finally did, I screwed up my courage and told my folks I wanted to see a psychiatrist.

They were stunned. "Only crazy people need psychiatrists," my mother said. "You're not crazy, you've just been reading too many depressing books. I told you not to read that Dostoyevsky and that *meshuggener* Kafka when you were fifteen." It didn't take me too long to prevail, however, and my parents, ever indulgent, agreed to pay for the psychiatrist, who was not cheap. Two visits per week, at ninety dollars per visit—a lot of money today, but a fortune in 1965.

Dr. Liebert's office was on East Eighty-seventh Street, in a high-rise with a doorman. He was a soft-spoken, balding man in his mid-thirties, calm, deliberative, obviously an intellectual. I enjoyed going to his office, sitting on the expensive leather couch, talking about my dreams and the events of my life. I was an enthusiastic patient, at least at first. Within days, maybe minutes, maybe even before he had met me, Dr. Liebert developed a theory about my character development's having been distorted by my "domineering [read Jewish?] mother" and my "distracted, absent father" who worked all the time. His analysis came straight out of the New York City Freudian casebook. The therapeutic method, I deduced, was to "transfer" my feelings about my father onto Dr. Liebert.

At one point Dr. Liebert told me that my worried mother had called him and asked, "Is he a homo?"

"I assured her you are not," he reported with a chuckle.

I went along with his theories for a few weeks, but there was something too pat there. Not knowing my parents, Dr. Liebert's one-line stereotypes missed the complexity of my relationship with them. Also, I had an intuitive sense that the problem lay elsewhere. I began to lose interest in his whole line of analysis as other involvements grew in my life.

On alternate afternoons to my shrink appointments—the irony didn't escape me—I'd take the subway uptown, in the opposite direction from Dr. Liebert's, to 145th Street. There I'd transfer onto a crosstown bus to Central Harlem, always the only white person aboard. I'd walk a few blocks to the ancient, dark tenement to meet the nine-year-old boy I was tutoring. Along with hundreds of other liberal white students at Columbia and Barnard, its sister college, I'd volunteered through a campus organization called the Citizenship Council to help an "underprivileged child." It was our do-gooder way of participating in the most important social movement of our time, the civil-rights movement.

Gary was bright and eager, yet he couldn't read. He was also behind in arithmetic. He wasn't getting any attention in school, which was no wonder because there were thirty-two kids in his class. We'd sit at his family's kitchen table and work together on his homework for the day. I'd prompt him as he read aloud to me. The one-on-one attention seemed to raise Gary's self-confidence; at least he said he was doing better in school. I wasn't so sure what we were accomplishing, but I tried my best, even though I had been given no training and had no idea what I was doing as a tutor.

Occasionally Gary and I would take a weekend field trip to some exotic place like Staten Island, riding on the ferry. He loved being away from the cramped apartment on West 148th Street. Both Gary's parents worked, so when he and his older sister got home from school, they'd let themselves in. Latchkey kids, they had strict orders not to go out on the street, lest they get into trouble. They were prisoners in the dark railroad flat, watching television all afternoon, every afternoon.

I couldn't help but compare my own childhood to Gary's. I had lived with my brother, David, eight years older than I, in a comfortable, ranch-style brick house in the suburbs, with a lawn and a backyard, surrounded by books from the library, with parents who drove us to music lessons, Hebrew school, Boy Scouts. We had always known—it was expected of us—that we'd do well in school and go to college. Our high school sent over 80 percent of its graduates to college; dropouts were rare.

Maplewood, New Jersey, sits on the western edge of Newark, where my maternal grandmother lived. When I was born, my family was living in a second-floor apartment near the Weequahic section of Newark, the Jewish enclave that novelist Philip Roth describes so brilliantly. Like many of Roth's families, we left Newark the next year for our first house, a little cracker box on the poorer side of Maplewood, just up Springfield Avenue from Irvington.

But my maternal grandmother, Ella Bass, with whom I was very close, stayed behind in a mixed working-class Italian-Irish-Jewish neighborhood called the Westside. From six in the morning to nine at night, she ran what was called at that time a candy store, where she sold penny candies, milk and soda, and dipped ice-cream cones from behind a marble soda fountain. When my father left the army right after World War II, he bought the six-unit tenement that the store was in. It was his first rental property.

As a little kid, I sat on the stoop of my grandmother's store, getting fat on all the candy and ice cream I could eat. Over time I watched the houses and tenement apartments of the surrounding blocks empty of Irish and Italian workers and Jewish storekeepers. They fled for the suburbs as black people from the Carolinas and even a few Puerto Ricans moved in. Every night I'd go home to our completely white town.

My grandmothers and my parents and the other relatives and all my parents' friends at the shul (synagogue) talked about the *schvartzes* (Yiddish for "blacks," but with the connotation of "niggers") who they said were destroying Newark. It was true that my grandmother's neighborhood was becoming unsafe. Grandma Bass would no longer walk alone to the Orthodox synagogue or to the stores a few blocks away, for fear of getting mugged. When I was about fourteen years old, she sold the store and moved into an apartment in a building owned by my father in a Jewish neighborhood of Elizabeth, the industrial city just south of Newark where my mom and dad had grown up and met as teenagers. My grandma died there in 1965, at the age of eighty-four. Four years later, after the Newark riots of 1967, the Jewish population of Newark would be close to zero, down from about a hundred thousand after World War II. That's my people, the ones who fled to the suburbs.

Tutoring Gary in Harlem was a lot like being back in Newark. But instead of taking the bus home from my grandma's to Maplewood, as I used to, I'd return to Columbia, my white island in the black and brown Manhattan sea. Every Monday morning a black woman in her mid-forties, a mother of two, would come up from Harlem to my dorm suite to clean the bathrooms. Her service was included in the room and board.

One day I argued with my shrink. "Look at your practice," I said. "You only take care of rich people. Just a few blocks from here, in Harlem, there are people suffering much worse than overprivileged white kids." As I said these words, I was completely aware of the ridiculousness of the situation, being myself one of those overprivileged kids, paying my parents' money to argue social ethics with a shrink.

"Oh, really?" Dr. Liebert replied, smirking at me in earnest. "Is it possible you're raising this right now because we've finally uncovered some painful feelings you don't want to look at?"

"Right, like poverty and racism," I said, and that was the end of my psychoanalysis.

One night during the second semester of my freshman year, I was reading the eighteenth-century social philosopher John Locke in my eighth-floor dorm room, depressed as usual and struggling to stay awake. On the wall in front of my desk was a print that I had bought at the Museum of Modern Art of a Joan Miró painting, *Inverted Personages.* I thought its spontaneous lines and bold colors signified my intellectual avant-gardism. I answered the door, and in walked David Gilbert, the chairman of Columbia's Independent Committee on Vietnam (ICV). I had seen Dave, a senior sociology major, standing by the ICV literature table, near the Sundial at the center of campus, debating supporters of the war. He always seemed to demolish them. I had also read his articles arguing against the war in Vietnam in the *Daily Spectator,* the Columbia College paper. Dave introduced himself and said he was out canvassing the dorms to find people interested in antiwar work on campus.

"What do you think about the war in Vietnam?" he asked.

"I'm pretty sure it's wrong, but I don't know too much about it," I replied.

Dave was direct and honest, also genuinely friendly. Quick to smile, he had a broad, open face, heavy eyebrows, and short, dark, curly hair. Over the course of the next few hours, we sat and talked. He didn't harangue me or make me feel stupid. He just quietly told me about himself and what he and others were doing. He came from a suburban Boston Jewish family, a background very much like my own. He'd been an Eagle Scout, and I had also been a Scout, though not attaining that level. Like me, Dave hadn't come from a left-wing background, but he had been inspired by the black college students who conducted the lunch-counter sit-ins in Greensboro, North Carolina. He described seeing Martin Luther King on television and thinking, "This is what it's like to be human, to be moral, to care about other people."

As a freshman at Columbia, in 1962, he had joined the campus chapter of CORE, the Congress of Racial Equality, one of the most militant civil-rights organizations. He had also, like me, gone into Harlem with the Citizenship Council program, tutoring a black kid in his home. His conclusion from that experience was very different from mine. "The goal," he said, "isn't to make black people more like you and me. It's that they take control of their own lives and their community. That's the radical position. The liberal position of superiority is condescending."

Dave told me that he'd had a breakthrough moment just a year before. While taking the train to Harlem, he'd read in the paper that the United States had just begun the sustained bombing of Vietnam. He became extremely upset, and it must have shown on his face when he got to the house of the kid he tutored. The child's mother asked him if he was okay.

He replied, "I can't believe it. Our government is bombing people on the other side of the globe for no good reason."

"Bombing people for no good reason, huh?" she said. "Must be colored people who live there."

"That was a complete revelation to me," he said. "She had never heard of Vietnam, but she naturally made the connection that I had failed to make, even though I'd been working on both fronts, peace and civil rights, for almost four years. I was blinded by calling our system a democracy with some faults, while she understood it as being in essence a racist and violent system."

"Have you read *The Autobiography of Malcolm X* yet?" he asked me.

"No, I've been too busy with all the reading for freshman CC and Humanities." Contemporary Civilization and Humanities were both required courses.

"Well, you'll learn a lot more about contemporary civilization from Malcolm than you will from reading Plato," he said. "When Malcolm made the connection between Third World peoples' struggles abroad to free themselves from U.S. imperialism and the black struggle at home, the CIA signed his death warrant."

I did know that Malcolm had been shot and killed at the Audubon Ballroom on Broadway in upper Manhattan just a year before. I made a mental note to read the book.

"I heard him speak at Barnard three days before he was shot," Dave went on, "probably the most formative experience of my life. He said that the division in the world isn't between black and white, it's between the oppressed—who are mainly people of color—and the oppressors—who are mainly white. He also said that white people can play a positive role by organizing within their own communities."

Dave seemed to be knowledgeable on these things in a way that I wasn't, but even so, I recognized a lot of myself in him. He struck a chord in me when he said, "Our country unjustly attacked Vietnam. I can't stand by and allow this to happen, like a good German." Any Jew alive at that time would have known he was referring to the great mass of Germans who, in their ignorance, in their denial, and especially in their silence, allowed the Nazis to do their work. "We didn't know," was their phony cry when asked after the war about the destruction of the Jews.

I didn't want to be a good German either.

"Assuming that what you say is true," I said, "and the war is morally wrong, so what? What can anyone do about it?"

"Well," he said, "we're part of a larger movement that will eventually end the war. People are working in different ways all over the country.

"We do what we can. In May, right before the summer vacation, we held an antiwar protest at the Naval ROTC graduation ceremony," he informed me, referring to the Reserve Officer Training Corps. "Twenty-five of us were there. We actually caused the ceremony to be canceled. The university called the city cops, and they got real nasty with us. They grabbed me and a few others and beat us with billy clubs." He paused. "You know what I learned?"

"No, what?"

"Don't ever wear a tie to a demonstration. That fucking cop choked me with my own tie," he told me, laughing.

There was something so charming, so smart and warm about this guy, that I told him I'd be at the next ICV meeting.

The meeting was held in an old classroom in Hamilton Hall, the main Columbia College instructional building, named after Alexander Hamilton, an alumnus from the days before the Revolutionary War when the school was known as King's College. The furniture consisted of one-piece wooden desk-and-chair units fixed in tight rows to the scuffed wood floors. A mixed group of undergraduates and grad students—predominantly guys, but a few Barnard women—were passionately debating which demand the Independent Committee should endorse in the upcoming Fifth Avenue Peace Parade, to be held in March 1966. Only a small number of people at the meeting supported the more limited demand for negotiations. Several on the other side, those who wanted the march to call for "immediate withdrawal," denounced them as "liberal opportunists." I wasn't sure what this meant, so I sat back and listened.

A red-faced junior named Harvey Blume stood up to argue for withdrawal. He was wearing thick, plastic-framed glasses, which he took off periodically to wipe, as his rhetoric and the atmosphere heated up. He delivered a fervent historical lecture.

"The Communist Party in this country failed because of its opportunist politics." There was that word again. "The party hid its intentions—working-class revolution—behind a screen of trade-unionism and 'progressive' electoral politics. They even hid their true identity as Communists. So when the government unleashed its anti-Communist campaign in the late forties, early fifties, the workers had no understanding of the Left's real aims, and they allowed the Communists in the unions to be thrown out. We have to call the war by its right name—imperialist aggression." That last line made people clap.

"The liberals say the war in Vietnam is a well-intentioned mistake. Bullshit! It was the liberals who started it. It's not the international Communist conspiracy we're fighting, it's the people of Vietnam themselves. Vietnam is part of the Cold War: It's about keeping a country from seceding from the empire. They can't stand the example of letting anyone be free; it might prove contagious. That's the real domino effect Kennedy and Johnson always talk about.

"By our speaking the truth about the war, by our saying that the U.S. has no legitimate right to be in Vietnam, people will begin to learn the true nature of imperialism. We'll be taking a step toward building a revolutionary movement and preventing future Vietnams. The ICV should vote for the 'immediate withdrawal' position against the opportunists and the other liberals in the Parade Committee." Much applause, shouts of "Right!"

These guys took themselves so seriously you'd think this was a debate of the workers' soviet of revolutionary Petrograd in 1917. But they were also mesmerizing, articulate, and burning with conviction. By comparison, my professors seemed tame and bloodless. In my

European history class, revolution was something that had happened in 1789 in France. But these zealous radicals in the ICV were talking about the class nature of the American system and about revolution happening right now, in this country as well as around the world.

After the ICV meetings, people would go to hang out at the West End Bar on Broadway or to an upperclassman's apartment. Over beers or a joint, I'd listen to discussions about China's Cultural Revolution, then just starting, and to Cuba's seven-year-old revolution. It was thrilling to be with these people who were tapping into something so much bigger than ourselves—something so grand, so *historic:* remaking the world.

There was this one guy, a freshman like me, named John Jacobs, or JJ, who got my attention because he was an animated madman. He talked in breathless whole paragraphs about Lenin and Mao and Marx in the nasal "dese" and "dose" accent of a working-class kid. Actually, JJ had been sent to an expensive prep school in Vermont by his wealthy Ridgefield, Connecticut, parents. He was almost my height, about five foot eleven, blond, thin but with very strong shoulders. His most prominent feature might have been his piercing blue eyes, which I later heard Barnard girls describe as "bedroom eyes."

JJ told me he had joined the anti-imperialist May 2 Movement right out of high school. M2M, as he called it, was a "mass front organization" for the Maoist Progressive Labor Party (PLP or PL). The party had let it run for less than two years, in which time it had grown by calling for militant resistance to the blockade of Cuba and urging men to resist the draft. JJ wasn't a member of PL himself, he told me, but he understood them.

In high-volume monologues that often shifted to yelling, JJ would hold forth while guzzling a beer and smoking a skinny joint. "In order to abolish colonialism and imperialism, we need to abolish capitalism, the root cause of war, domination, class and race exploitation. When that happens, we can substitute a humane, rational economic system, socialism. The majority of the people of the world will benefit, war will be eradicated, and history can then begin, as Marx himself said somewhere. An end to exploitation! Human beings will be liberated from servitude and toil for the first time in human history. What a wonderful era we're in!"

JJ was among the large number of my new antiwar activist friends who came from left-wing—that is, socialist or Communist families. It was from these "red-diaper babies" that I learned about "the struggle" that had been going on for generations. I felt lucky to be among them, because they seemed to embody an idealism that had passed my family by. My immigrant grandparents on both sides were not political, nor were my parents. They were pragmatists, too busy making a living to become involved in dreams of a better world. My father had seen friends he grew up with in Elizabeth become Communists in the thirties, then lose their jobs as teachers or engineers when the McCarthy repression was unleashed in the early fifties. He had often told me that his strategy for survival was to keep his head down.

The Great Depression of the thirties was the defining event in both my parents' lives. They came of age when there were no jobs to be had, and no money. My father earned two pensions by the time he was fifty, first as a lieutenant colonel in the U.S. Army Reserves and later as a civilian working for the Army & Air Force Exchange Service, the PXs. He and my mom then built a business owning and managing apartment houses. Their lifelong goal, which they successfully reached, was the American dream of financial security. All these red-diaper babies talking about socialism challenged me to want something loftier: a utopia of freedom and justice and peace.

One other thrilling aspect of all this talk was that it gave us, the students with our brilliant ideas, a role to play. By agitating on campuses, we could have a hand in building the mass movement that would eventually change this country and the world.

Plus, having served my sentence as a loner, a bookworm, a nerd in high school, I had finally found my gang.

HEY, HEY, LBJ, HOW MANY KIDS DID YOU KILL TODAY?" Thousands of us screamed that slogan on a warm spring Saturday, March 26, 1966, as we marched down Fifth Avenue from Ninety-first Street to Seventy-second. Our Columbia contingent was several hundred strong, brought out by the Columbia Independent Committee's leafleting and dorm canvassing. This was the first time I'd heard the chant, but I would yell it thousands of times in the coming years. Dead Vietnamese were never counted, but the American numbers were quite precise: At the end of 1965, 636 U.S. military personnel were listed as dead; one year later that number had multiplied by a factor of ten to 6,644.

I later read in the newspaper that the crowd at my first march was fifty thousand. Where I was, it was predominantly students, but I was delighted to see large numbers of adults, including parents with kids in strollers, workers marching behind union banners, elderly people, and, significantly, contingents of veterans from World War II and the Korean War wearing their knife-edged "overseas caps." These older demonstrators seemed to me to represent typical middle-class Americans—prosperous, respectable, white.

There was one startling exception: A large contingent of several hundred black people marched down Fifth Avenue from Harlem behind a large banner reading, THE VIETCONG NEVER CALLED ME A NIGGER! It was a quote attributed to world heavyweight boxing champion Muhammad Ali, who had refused induction into the military. A year later, in complete violation of his constitutional rights, he would be stripped of his title as punishment.

My group from Columbia passed by a bunch of counterdemonstrators holding up signs and screaming at us, "VICTORY IN VIETNAM!" and "COMMIE TRAITORS GO BACK TO RUSSIA!" and "SUPPORT OUR BOYS IN VIETNAM!" I have etched in my memory the hate-contorted face of one young guy about my own age who tried to push through the parade marshals to get at us.

My crowd of elite Columbia students struck back in the way we knew best—with words. Michael Neumann, my philosophy-student friend who would become my roommate two years later, started the snotty chant, "READ BOOKS! READ BOOKS!"

The parade ended at the Mall in Central Park, which was already overflowing with thousands of people chanting, "WHAT DO YOU WANT? PEACE!!! WHEN DO YOU WANT IT? NOW!!!" Most of the banners and signs held by the crowd read, BRING THEM HOME NOW! I was glad to see that the "radical" position had won out.

I roared, along with most of the rest of the crowd, when a speaker asked the rhetorical question, "If you saw a rape in progress, would you call for negotiations or for immediate withdrawal?"

As I was leaving the park with my friends, chatting and laughing, I caught a sudden glimpse of my shrink, Dr. Liebert, passing near me. Our eyes met for just a flash; we didn't acknowledge each other, but at that moment, inexplicably, a huge lump of sadness welled up from deep inside me. I felt like crying.

At the end of my first year at Columbia, in the spring of 1966, the Independent Committee on Vietnam voted to convert itself into a chapter of Students for a Democratic Society, SDS, a national organization. I vaguely understood that the change had something to do with creating a radical student organization at Columbia that would address more issues than just the war. It would unite such issues as racism, control over our own lives at the university, and imperialism as the underlying cause of war. "Radical," I quickly picked up, meant getting at the root of things.

John Fuerst, skinny, long-faced, and hyper-intense, became chairman of the new SDS chapter—David Gilbert was graduating and going downtown to study sociology at the New School. John gave me a copy of its founding document, the Port Huron Statement, which had been written in 1962 by a small group of activist students led by Tom Hayden. It was more than fifty pages long, and it challenged basic status quo assumptions about the country and the world.

I was astonished by Hayden and the other authors' intellectual breadth: In a single tour de force, they addressed politics, economics, the Cold War and anti-Communism, the nuclear arms race, decolonization around the world, and the civil-rights movement at home. They asked fundamental and important questions: How could poverty exist in the midst of such plenty? Why did the war system persist when we had the possibility of making peace? Ambitious and imaginative at the same time, the young writers proposed the realignment of U.S. political parties into true conservative and liberal, the revitalization of the public economic sector, the elimination of poverty and the spreading of civil rights, disarmament, and an end to the Cold War.

I was attracted to their intelligence, yes, but it was the idealism of the writers of the Port Huron Statement that pulled me in: "We would replace power rooted in possession, privilege, or circumstance by power and uniqueness rooted in love, reflectiveness, reason, and creativity." The "participatory democracy" they proposed would allow each individual citizen to make the decisions affecting his or her life. Plus, the Port Huron Statement gave students an important role in the development of a "New Left" in this country, as the visionaries and activists for the new society.

Sign me up!

A few days later, I paid John my five-dollar dues. On the spot he filled out my membership card, which had LET THE PEOPLE DECIDE! printed at the top. I was now a card-carrying member of SDS. I also began wearing a little lapel button, about the size of a dime, white with the simple brown logo *sds* in lowercase. Sometimes that button would appear on my blue blazer. More often I'd substitute jeans, work shirt, and work boots, my comfortable new radical getup. By the end of that year, my preppy clothes were hanging permanently in the closet.

Many of the drafters of the Port Huron Statement were young white people like Tom Hayden, its principal author, fresh from the civil-rights battle lines in the South, where they had worked with the Student Nonviolent Coordinating Committee, SNCC. In the intervening four years, they had ambitiously sought to build something similar in the North, an "interracial movement of the poor," by organizing in working-class neighborhoods in northern industrial cities like Chicago, Cleveland, and Newark around jobs and income and housing conditions and welfare. SDS was often seen as a sister organization to SNCC: Both organizations considered participatory democracy a method as well as a goal. The first generation of SDS leaders had picked up the concept from SNCC in the South.

But the U.S. government had been gradually escalating the war in Vietnam during these years: At the end of 1964, there were 23,300 "advisers" stationed there. That number multiplied to 184,300 at the end of 1965, the year I turned eighteen and went down to my draft board to register, as U.S. ground forces went into direct combat, something three presidents had vowed would never happen. By the end of 1966, we had almost 385,000 GIs stationed in Vietnam, plus many thousands more in support positions around the world. Twenty-five thousand young men per month were drafted. Had I not been in school, I would have been drafted and sent to Vietnam.

As this escalation continued, SDS gradually shifted its focus from community organizing around economic issues toward Vietnam. The organization had called the first national antiwar demonstration in Washington on April 17, 1965. The marines had just landed at Da Nang, and the air force had begun its years-long Operation Rolling Thunder in North Vietnam to, as one general put it, "bomb Vietnam back to the Stone Age." Ten thousand people were expected; twenty-five thousand showed up.

A year and a half later, I picked up an SDS pamphlet that contained the transcript of SDS president Paul Potter's historic speech ending that first antiwar rally in front of the Washington Monument. As I read, I realized that he had captured my own growing hurt and sadness about the war:

> The incredible war in Vietnam has provided the razor, the terrifying sharp cutting edge that has finally severed the last ves-

> tige of illusion that morality and democracy are the guiding principles of American foreign policy. . . .That is a terrible and bitter insight for people who grew up as we did—and our revulsion at that insight, our refusal to accept it as inevitable or necessary, is one of the reasons that so many people have come here today.

By "people who grew up as we did," Potter was referring to me, the son of a World War II vet, who had spent my whole childhood believing that my country was a force for good in the world.

Potter then described the destruction and dictatorship our government was perpetrating on Vietnam. The U.S. goal was obviously not self-determination and freedom. At a midpoint he asked the questions that made the speech famous:

> What kind of system is it that justifies the United States or any country seizing the destinies of the Vietnamese people and using them callously for its own purpose? What kind of system is it that disenfranchises people in the South, leaves millions upon millions of people throughout the country impoverished and excluded from the mainstream and promise of American society. . . ?
>
> We must name that system. We must name it, describe it, analyze it, understand it, and change it.

At this point people in the crowd were screaming in response, "CAPITALISM!" and "IMPERIALISM!" Potter later wrote that he declined to use the word "capitalism" not out of opportunism but because it had become "for me and my generation an inadequate description of the evils of America—a hollow, dead word tied to the thirties."

Prophetic, Potter ended his speech by describing the movement we must build to end the war:

> . . .a movement that understands Vietnam in all its horror as but a symptom of a deeper malaise, . . .a movement that will build on the new and creative forms of protest that are beginning to emerge, such as the teach-in, and extend their efforts and intensify them; that we will build a movement that will find ways to support the increasing numbers of young men who are unwilling to and will not fight in Vietnam; a movement that will not tolerate the escalation or prolongation of this war but will, if necessary, respond to the administration war effort with massive civil disobedience all over the country, that will wrench the country into a confrontation with the issues of the war. . . .

At the time—and still today—the moral and intellectual clarity of this speech sends shivers up my spine. It reminds me why it was such a thrill to be a member of SDS.

In my classes at Columbia, I was learning about the glories of Western civilization—the philosophers and literature that produced the Enlightenment, the rise of democracy, the idea of individual liberty. A few professors taught radical philosophy, such as the teachings of Hegel, Marx, and the contemporary Herbert Marcuse, but it was an abstract radicalism, unrelated to the current world. It seemed to me that most members of the faculty, even the renowned ones with leftist reputations, were not at all concerned with what the Left could accomplish today.

At SDS meetings I would listen to passionate debates by upperclassmen and graduate students about China's Cultural Revolution, the Cuban revolution, the nature of American class society, this country's role in blocking national liberation movements, including those of minorities such as blacks inside this country. They all agreed on one solution, Marxist revolution. I wanted to be like all these brilliant and burning young men—Michael Josefowicz, Harvey Blume, Howie Machtinger, David Gilbert, John Fuerst, Michael Klare—to understand the world as they did. I never once heard Vietnam mentioned in a class, while outside of class Vietnam was everything.

Along with others in SDS, I was reading the works of Australian journalist Wilfred Burchett, who described how the United States had assumed France's colonial role; how we had placed a handpicked dictator, Diem, in power and kept him there with massive bombing from the air and an enormous repressive puppet army. I also learned from Burchett, as well as from two publications, the scholarly *Viet Report* and the radical weekly *National Guardian*, that the Vietcong were a popular insurgency, guerrillas surviving among the people "like fish in the sea." The American strategy was to dry up the ocean—that is, remove the peasants from the countryside. The U.S. military used a host of brutal tactics, ranging from herding people into guarded concentration camps called "strategic hamlets" to destroying peasants' crops with herbicides like Agent Orange to simply murdering them and destroying their villages with B-52 carpet bombing or artillery fire. As a result of our terror attack on the countryside of Vietnam, we generated 15 million refugees, most of whom went to the cities seeking safety and work.

I learned that the American military was an occupying army in Vietnam, and as part of such, no American soldier could ever feel secure. A kid carrying a cigarette pack filled with explosives could kill you as easily as could an armed guerrilla with a rifle. In this environment the rules of warfare made little sense, and all Vietnamese became the enemy. In 1969 the country would hear about the massacre at My Lai, which was not at all an isolated incident, but as early as 1966 I was reading about murder and torture by American troops in Vietnam. Given the nature of the war as a military occupation of a civilian population that did not want our soldiers there in the first place, such brutality was inevitable.

And what was the war for? An SDS pamphlet by David Gilbert, entitled *U.S. Imperialism,* explained that throughout the Third World, people were attempting to throw off the yoke of American and European control. That accounted for the dozens of U.S. military "interventions" around the world in the twentieth century. Revolts in such places as Cuba, Guatemala, Iran, Algeria, China, and Vietnam were "national liberation struggles"; these were the natural enemy of imperialism.

For me, only the radical analysis explained what was happening in the world. I wanted to place myself on the side of freedom, not empire.

Vietnam touched a wound inside me. The Holocaust had been a fact of my entire childhood, though I was born two years after World War II ended. In my home, as in millions of Jewish homes, "Hitler" was the name for Absolute Evil. As a kid I listened to two of my father's cousins describe how they had survived the concentration camps, and the whole while I was sneaking glances at the numbers tattooed on their arms. The suffering in their voices and their eyes told just part of the story; the rest I could only imagine. I was terrified, appalled, spellbound, reading the diary of Anne Frank or looking at *Life* magazine's images of Jewish families being rounded up, as well as the pictures taken when the camps were finally liberated.

With the solipsism of a child, I saw myself among the dead. Over and over I pondered what would have happened to me and my family if my father hadn't emigrated in 1917, or if my mother's family hadn't come to this country in 1911? Why was I allowed to be born while so many millions of other Jewish kids perished?

The Holocaust brought me to the knowledge that evil exists and that it is associated with racism; that's what Nazi anti-Semitism was. Growing up watching the civil-rights movement in this country and then learning about Vietnam, I saw evil again. Only this time it wasn't the sick, repressed Germans with their little brush-mustached dictator. It was us, the Americans, the most democratic, the most productive, the most egalitarian people ever to have graced the earth. We were responsible for these horrible atrocities.

Here I was, stuck with the knowledge of the evil of the American empire, the racism, the violence, the torture, the corruption. What would I do with this knowledge? Would I have the courage not to be a Good American and turn away?

Right before the Thanksgiving break in 1966, the SDS chapter decided to conduct a three-day fast and vigil for peace. The goal was to draw people's attention to the immorality of the war. The idea of a fast was proposed by Ted Kaptchuk, who next year would be elected chairman of SDS. Ted was a thoughtful, soft-spoken religion major who was sensitive to the importance of moral statements. The son of leftists, he was looking for a more spiritual path.

I fasted the first day, sitting out on the massive steps in front of Low Library with the other vigil keepers between classes. We attracted the attention of the campus press and passersby. My problem was that three days is a long time: I love to eat and always have. So the last two days my moral steadfastness wavered, and I snuck a piece of fruit here, a sandwich there, when no one was looking. I didn't mention it to any of my fellow fasters on the steps.

The night the fast was over, I threw a "break-the-fast party" at my apartment on 116th Street, right across Broadway from Columbia. As a second-year student I had moved out of the dorms as soon as I could and into a big old Upper West Side apartment with three other guys. I warmed up my mother's frozen chicken soup with matzo balls and made my own pot of spaghetti and meat sauce. I was following an old cultural tradition: After Yom Kippur, the Day of Atonement, on which Jews fast, it's traditional to have a special break-the-fast meal. There was family precedent, too: My father had been a mess officer in the army, and I, perhaps genetically programmed, enjoyed feeding people.

Toward the end of the party, I admitted that I hadn't fasted all three days. "Look," I said, "we're not guilty for the war in Vietnam. Lyndon Johnson, Dean Rusk, and Robert McNamara should be fast-ing, not us. They're the guilty ones." For some reason my position seemed like rationalization to my friends, though it made perfect sense to me. I guess I just didn't understand spiritual statements.

Over the next months, my involvement in SDS continued to grow, including participating in direct action. Not long ago I found three letters from March 1967, in a box of stuff my mother had saved for forty years. The first was a letter I typed on onionskin paper. It opened:

> Dear Parents,
>
> Hello. It's your son writing. I participated in a little hoo-hah with the Naval ROTC here on Monday. The whole thing got me a letter from the Dean of the College warning me about doing bad things like I did.

I went on to tell them that I had taken part in a sit-in at a Naval ROTC class. I also explained that we did it in order to strike a blow against the war, to try to get the program thrown out of Columbia.

On official Columbia University letterhead from the Office of the Dean, signed by Dean David B. Truman, was a letter, dated March 6, 1967, and addressed to me:

> Dear Mr. Rudd:
>
> It has come to my attention that you, along with some other students in the college, have visited in the past few days certain of the Naval Science classes without the permission of the instructor and have declined to leave except when requested to do so by Dean Platt.

Dean Truman then lectured me about the inappropriateness of attending a class without permission and suggesting that if I had any criticisms of university policy, I might submit comments directly to him.

The third letter was a carbon copy of my reply, which my mother saved.

> Dear Dean Truman,
>
> Today I received a letter from you concerning my having attended the Naval Science class on Monday, March 6, a letter which warned me against future violations of the rules and customs of the faculty. . . . Indeed, my decision to violate College rules was made only after a long process of evaluation of my personal goals and of my relationship to society.

In the rest of the four-page, single-spaced letter, I gleefully and pedantically lectured Dean Truman in return about the difference between "functional rationality" and "substantial rationality," citing numerous sources from readings in my sociology class. Functional rationality, I explained to him, was fitting means to ends, like Defense Secretary McNamara's engineering of the war; substantial rationality meant opposing the evil inherent in the destruction of human life, as in the Nazi massacre of the Jews in Europe or the current American murder of Vietnamese. That was why I felt compelled to take action.

I wrote, "I have extreme respect for the law; I am by no means an anarchist. Yet I respect the law less than I respect morality; I will have to break the law if upholding it means the continuation of brutality."

Dean Truman didn't respond to my letter.

2

Love and War

In April 1967, Columbia SDSers read in the *Spectator* that U.S. Marine recruiters would be on campus for two days, looking for prospective officers. We took this as a direct provocation: The Vietnam War was coming to Morningside Heights.

SDS had been exposing and challenging the university's involvement with the war since the previous fall. Along with the sit-in at the Naval ROTC class that I participated in, we had led a successful referendum campaign to keep Columbia College from sending students' rank in class to local draft boards. Presumably, they would use the information to draft the poorer students. In the atmosphere of urgency before the referendum, I had ceased attending classes to help organize forums and dorm meetings. Though we won the referendum and the faculty subsequently voted to withhold the rankings, the majority of the students were driven more by self-interest—staying out of the army—rather than by a principled opposition to the government's policies in Vietnam.

In February a splinter group of SDS, people associated with the Maoist Progressive Labor Party (PL) had sat in to block recruiters from the Central Intelligence Agency from interviewing potential employees. The recruiters left, but sixteen people received warning letters from the university for participating. My friend JJ joined the sit-in, but I stayed with the majority of the SDS chapter that had voted against disrupting the recruiting. I was swayed by the arguments of Ted Kaptchuk, the chapter chairman, and his vice chairman, Ted Gold, who felt that the disruption would polarize the campus against us because there was not a general understanding of why recruiting for the CIA was wrong. We were scared of being politically isolated from other students.

Still, the Progressive Labor Party hard-core faction and its supporters ignored the vote and staged the sit-in. A bitter debate around the issue of disruption broke out on campus among faculty and students. Most of the faculty, both liberals and conservatives, believed in the "neutral" and "objective" character of the university. They did not want to see its normal functioning disturbed because of "external" political issues. Further, several senior faculty members were refugees from Nazi Germany. They identified SDS's antiwar activities with the actions of Nazi students at German universities in the thirties. This was a frightening image, even to us in SDS. "What if every small group had the power to silence whomever they wanted—such as you?" asked the old liberals. "Isn't there an absolute right to free speech?"

I found myself in many heated arguments with my professors and other students about whether Columbia was participating in the war by allowing recruiting. "Suppose the Nazi SS came to a German university to recruit. Would you be silent and just let them?" I'd ask. "The Nuremberg Trials [of Nazi leaders after the end of World War II] established that it is the legal and moral responsibility of the individual not to comply with orders that constitute war crimes." I also argued that interruption of the "normal" recruiting function at the university—which appears to be free speech—was a much smaller evil than the enormous one that recruitment served, the war.

After a long spring of debate, the issues of complicity and disruption were coming back again, along with the U.S. Marines. The first day their recruiters appeared, SDS mustered around two hundred students to "confront" them. Actually, we just milled around in the dorm lobby where the marines had a table. Our exact intention was ambiguous: Even we didn't know what we would do. Opposing us was a group of about fifty right-wing and ROTC students—whom we called "jocks," whether they were athletes or not—protecting the marines by standing in front of them with their arms folded.

The two sides started pushing and shoving in the small lobby. "Fucking jock," some of our guys taunted. "Commie puke," we got in return. An SDSer was hit with a punch; we got our licks in, too. A college dean was urging on the jocks from the sidelines. I stayed with the SDS front line, facing off with a jock. We glared at each other, neither of us wanting to throw the first punch. Deep down I knew I was no fighter, but I would try my best to defend the cause. Just as the tension was reaching its peak, the campus police waded into the crowd to separate us. An official announced that the marine recruiting was over for the day. We had won.

About thirty of us SDSers went to the back room of the West End Bar to lick our wounds, plan for the next day, and celebrate our victory with a beer. We decided to mobilize everyone we could for the second day's confrontation: Tomorrow we would hold a peaceful demonstration against the marines and against the university's support for their recruiting.

That night we called everyone on our phone lists, put out leaflets to the dorms and the campus, and made announcements on the university radio station, WKCR. We also organized ourselves into a contingent of marshals, with white armbands, to keep order. I was a marshal.

The next day I was astounded to count more than eight hundred students marching in opposition to the marines. It was the largest antiwar demonstration yet at Columbia. The hourlong picket line that filled the small quadrangle between John Jay Residence Hall, where the recruiters were, and Hamilton Hall, the main college classroom building, felt like a celebration. We sang freedom songs from the civil-rights movement: "Just like a tree that's planted by the wa-aw-ter, / We shall not be moved." We chanted, "What do you want? Marines out! When do you want it? Now!"

We had never before mobilized an antiwar demonstration on campus with that many people. Our confrontational tactics had worked: We'd forced people to choose sides and take a stand against the war. It was a lesson we wouldn't forget.

The broader off-campus antiwar movement was burgeoning, too, in 1967, corresponding to the increasing numbers of U.S. troops in Vietnam, the media reporting of American casualties and gruesome images of the carnage, and the growing distrust of LBJ and his war managers.

The best indicator was the April 15, 1967, Spring Mobilization to End the War in Vietnam, which consisted of two simultaneous demonstrations, one in San Francisco and the other in New York City. About seventy-five thousand people turned out on the West Coast, but in New York, where I was, the numbers were uncountable. The official police estimate, dutifully reported in the press, was a hundred thousand people. The Mobilization organizers claimed four hundred thousand.

The starting point for the march was the Sheep Meadow, a vast open field in Central Park that often holds a hundred thousand people for public events. It was still jammed hours after the march to the United Nations began. I had walked downtown from Morningside Heights with hundreds of Columbia students. Along the way we joined with a column of several thousand Harlem residents. We also encountered a "revolutionary contingent" carrying blue and red National Liberation Front (NLF) of South Vietnam flags with a bright yellow star in the middle, plus banners proclaiming, VICTORY TO THE NLF! and red flags signifying revolution. I felt a secret thrill: Here were people declaring in public what I only dared to say in private.

The crowd was so large that the Columbia marchers never did make it to the rally at UN Plaza, where the featured speaker, Martin Luther King Jr., openly denounced the war after years of silence. The plaza was filled. Held up on a narrow crosstown street, prevented from going farther toward the UN by the gridlock caused by the thousands of marchers, I looked up to see a group of nuns waving at us from a second-floor balcony of their fancy Upper East Side convent. Everyone in New York seemed to be against the war.

I later learned that the day's events had begun that morning with several hundred young men burning their draft cards in a "We Won't Go!" protest organized by Cornell SDS. Though I applauded them, feeling that every act of resistance to the war was useful, that particular action wasn't for me. I had decided to keep my student deferment so I could continue to work against the war, rather than be tied up in a court case for draft-law violation. I felt no moral qualms about using whatever privileges I had.

That summer—dubbed "Vietnam Summer" as an echo of the 1964 Mississippi Freedom Summer—I volunteered to organize a draft-resistance project at a largely black and Latino working-class high school in the Bronx. Our results were minimal, if any. Ever good at rationalization, I saw no contradiction between the student draft deferment in my pocket and the resistance stance I was asking less privileged working-class high-school kids to take.

That summer of 1967, I fell in love with Sue LeGrand, a Barnard student a year older than I was. We were introduced at the student center by mutual friends in SDS, and I asked her what she was doing that night. She replied that she was free. I said, "Great! Can you pass out some leaflets at the corner of Broadway and 110th Street?"

To me, Sue had the exotic air of the American Midwest about her.

A slight brunette with a sweet face and curled hair, she was perky and irreverent like the female lead in a 1940s movie. Everything about her drove me wild; her broad Middle American accent, with a tinge of the South, pronouncing "pin" for "pen," was a total sexual turn-on. She was my very own American shiksa.

You'd have to have grown up in New Jersey in a tight, insular, Jewish family and community to understand this. Or you might try reading *Portnoy's Complaint* by Philip Roth. Shiksas—gentile girls—were the forbidden object of desire for me (and generations of Jewish men and boys) since the moment my parents had kept me, at the age of twelve, from accepting the invitation of the cute red-haired girl who sat in front of me in homeroom to attend her church's hayride. Sue was a nonneurotic American young woman—the daughter of a Catholic and a Baptist, for God's sake—and she liked me! She was smart, funny, and committed to fighting against the war. Add the Beatles, some grass, and you had this Jewish boy's fantasy. It probably worked for her the same way in reverse: We Jersey Jew boys can be pretty exotic, depending on your point of view. Or maybe she just liked me.

The Beatles released *Sgt. Pepper's Lonely Hearts Club Band* in America on June 2, 1967, which just happened to be my twentieth birthday. Listening to it over and over, Sue and I would lie in bed together, stoned on grass, and hear, "It was twenty years ago today. . ." and just laugh and laugh. Or we'd make slow, goofy love to the psychedelic strains of "A Day in the Life"—"He blew his mind out in a car. / He didn't notice that the light had changed"—coming at the moment of orchestral abandon. Later Sue would refer to that summer as "the golden summer." I was her first boyfriend; our first kiss was on the George Washington Bridge, high over the Hudson River that divides Manhattan from New Jersey.

I had been "sexually active," as Ann Landers refers to fucking, since I was fifteen. My first girlfriend was not a girl at all, but a married woman, Karen, age twenty-seven, who had a kid. I'd go over to her house after school, and we'd make love quickly, before her daughter came home an hour later. It was dynamite for my ego: Within just a few years of puberty, I'd achieved my masturbatory fantasy of being inside a woman. Sex was a great thrilling mystery to me and remains so to this day.

Fear of pregnancy, the dark side of my adolescent dream, had been eliminated because the Pill had just come onto the market. The setup was perfect, except for the small matter of the potentially jealous husband, who never did find out. I didn't feel the least bit guilty toward the husband, whom I never met, but I did feel funny because I knew I really didn't love Karen. I had read in novels about sex without love, but I was still surprised at how easily the two are disconnected. We continued off and on for almost two years, until I discovered girls my own age.

I didn't play sports as a kid or in high school. At the early age of five, my favorite book was *The Story of Ferdinand the Bull,* and I must have soaked up its lesson: Why fight when you can sit quietly and smell the flowers? I was overweight and soft, a lonely, nerdy kid, but I felt like somebody special when I was in bed with a woman. While other teenage boys were out proving themselves on the football field or the basketball court, I had better things to do.

My senior year of high school, I found an even better situation. I fell in love with Liliana, who was the school intellectual, thin and dark-complected, with sad brown eyes, long brown hair that made her the embodiment of a beatnik chick. The daughter of German-Jewish immigrants, intellectuals who had fled Hitler in the late thirties, Liliana introduced me to psychology, philosophy, modern art, literature, films, left-wing politics. I adored her for the world she opened up to me, which I hoped would be mine once I escaped Maplewood. Sex was a big part of that world, too: Her parents were Freudians and Marxists, and so, opposing all forms of repression, they condoned teenage sexuality. In practice this meant we'd walk to her house every day after school and jump into bed until her folks came home from work. What an arrangement.

On October 21, 1967, the Pentagon, the seat of American military power, the largest office building in the world, was almost levitated off its foundation by the energy of thousands of antiwar demonstrators surrounding it, flowers held out to rifles as in the famous pictures documenting that event. I missed it because I had been seduced into agreeing with the position coming out of the SDS National Office (NO) in Chicago to the effect that national demonstrations were a waste of time and sapped energy from more important local organizing. The leadership of the Columbia chapter, too, had argued against participating, and I had sided with them. Of course, at the last minute the NO had changed their minds and endorsed the demonstration, plus many SDS members and key organizers went to the Pentagon, including my friend Jeff Jones from the New York Regional Office, but I was stuck back in New York City thinking, incredibly, that the Pentagon demonstration was a waste of time.

The next day I read the stunning headline in the *New York Times*, GUARDS REPULSE WAR PROTESTORS AT PENTAGON, followed by the first line, "Thousands of demonstrators stormed the Pentagon today." I instantly realized how stupid the debate had been. This protest had taken the antiwar movement to a whole new level. National demonstrations feed local organizing and vice versa. I felt like kicking myself.

The week before the Pentagon, there had been five days of Stop the Draft Week demonstrations at the Oakland, California, induction center—alternating peaceful sit-downs with violent hit-and-run actions. Some ten thousand SDS-led demonstrators had outwitted the police and forced the induction center to close for a few symbolic hours.

Next to the *Times* article was a photo of a vast crowd surrounding the Pentagon. I suddenly had a vision of what needed to be done: The whole country should be forcibly awakened from its sleepy acquiescence to the government and its lies about Vietnam. Peaceful demonstrations were much too tame, too accepting of the mass murderers in

Washington who planned and approved the war. The Pentagon demonstration, with its confrontational tactics, had been an enormous success—the time had come to fight fire with fire.

I had been blinded by intellectual formulas about how radical politics should work. "Educate, build the base!" was the mantra of the Columbia SDS chapter. While Berkeley students were fighting cops in Oakland and young people from all over were laying siege to the Pentagon, I had been involved in sterile debates at Columbia SDS meetings about "correct" organizing strategy, whatever that was.

A week or so later, Jeff Jones of the New York City Regional Office of SDS, a happy-go-lucky long-haired blond surfer from Southern California improbably misplaced in Manhattan, told me that Secretary of State Dean Rusk would be given an award by the Foreign Policy Association, a true establishment organization. The presentation would take place at a black-tie dinner at the New York Hilton on Sixth Avenue. Rusk was one of the main architects and apologists of the war. I immediately saw a chance to redeem myself and the Columbia chapter.

The large coalition known as Vietnam Peace Parade Committee called a legal, peaceful demonstration to be held across the street from the Hilton. New York Regional SDS put out a leaflet that hinted at something different:

EMBROIL THE NEW YORK HILTON (6TH AVE. AND 53RD ST.)
REVOLUTION BEGINS: NOV. 14, 5-5:30 PM.

Before the demonstration we distributed plastic bags of cow blood to remind onlookers of the blood being spilled in Vietnam. The police had announced an official protest area about two blocks from the entrance to the Hilton, where no one would see the demonstration, but we wanted no part of police restrictions. Our goal was to disrupt the banquet from outside by stopping traffic.

For several joyous hours that warm fall evening, I ran with my "affinity group," including Sue, through the midtown streets, stopping traffic, banging on limos, dousing them with cow's blood, and grappling with the police, who were out in force. Screaming and yelling, we actually did block the streets. We had little concern for bystanders inconvenienced by our demonstration; in fact, our goal was to disrupt people's normal lives in order to compel them to consider the war. What they were probably considering was how much they despised us for inconveniencing their commute, but that thought didn't occur to me until years later.

I was having a great time, running and shouting and beating on cars. After all the discussions and arguments and peaceful protests, this was actually doing something physical against the "masters of war." But then suddenly I found myself pushed against a newsstand by a plainclothes cop with a walkie-talkie. Right behind him were four uniformed cops whom he directed to grab and handcuff me. I was caught, but I didn't care. As I waited to be driven off, I looked out from the backseat of a squad car at Jeff Jones and Sue, standing on the sidewalk, laughing, hugging, waving bye-bye to me. Jeff had held back at the demonstration because he was set to fly off to Cambodia the next day on the first leg of a trip to North Vietnam to meet "the enemy."

This was my first time in jail—the Tombs, the House of Detention for Men, at 100 Centre Street in lower Manhattan. I sat in a grimy holding cell with dozens of other young people picked up off the streets that evening. As newcomers arrived, they told us that they had formed roving bands that had rampaged all the way down to Times Square, stopping traffic and breaking windows in hit-and-run actions for several more hours. We had successfully disrupted much of midtown to protest Dean Rusk and the war.

In that holding cell was a guy named Abbie Hoffman, who was already a well-known political street organizer as well as a civil-rights veteran. He would go on to found the Yippies and then become a fugitive. Abbie's wild curly hair and enormous Jewish nose dominated his face, but it was his infectious humor and spirit that attracted people. I once heard Abbie described as the "pied piper of the counterculture." That night in jail, we sang. He and I knew the same folk songs, including, I remember, "The Salvation Army Song": "Throw a nickel on the drum, save another drunken bum. / Throw a nickel on the drum and you'll be saved." We sang all sorts of leftist anthems popular in folk circles at the time: Woody Guthrie songs from the Depression, like "Dough Re Mi"; Joe Hill's Wobbly anthem, "Pie in the Sky"; and Pete Seeger's version of the song from the Spanish Civil War "Viva la Quince Brigada." With unrelenting energy—in retrospect, manic—Abbie rapped nonstop between songs to the younger cellmates about the need for resistance, for building a movement. He probably didn't remember me from that encounter, but I sure remembered him.

Very early the next morning, I was arraigned in night court along with two other Columbia SDSers, Ron Carver and Ted Gold, the vice chairman of the chapter, on the charge of incitement to riot. Then we were released. I was proud of the charge: I thought incitement to riot was better than mere disorderly conduct or some such petty offense. That first arrest was a badge of courage.

When my case came up a few weeks later, my volunteer lawyer told me to plead guilty to the lesser charge of disorderly conduct. He had neither the desire nor the ability to wage a political defense, which would have involved arguing the necessity to break the law in the interest of stopping a greater evil. The judge gave me a lecture on courtroom decorum—"Don't sling your coat over your shoulder, this isn't a supermarket"—and on what he called, "legitimate and lawful protest." Then I was unconditionally released.

My lenient sentence was only just, I thought. Dean Rusk and the members of the Foreign Policy Association, not the protesters, were the ones who should have been arrested for war crimes,. As for the question of "legitimate and lawful protest," I was convinced by now that we needed more, not less, militant disruption of business as usual, like the Pentagon and Rusk demonstrations, in order to grow the movement and stop the war.

3

Action Faction

About a month after the Hilton demonstration, I got a call from the SDS National Office inviting me to join a delegation of students going to Cuba. This was the first SDS group ever to go to Cuba, and the first North American student trip since 1964. Travel to Cuba was illegal, according to the dictates of the U.S. government. I had been reading and thinking about Cuba and socialism for so long that I jumped at the chance to make the trip. We were supposed to fly there via Mexico City.

Fortunately, I was doing well enough in my schoolwork that I was able to arrange with my professors to take off the month of February. They were as curious as I was about what I'd find there, because very few Americans had been to Cuba since Batista's fall in 1959.

I had some money saved from a summer job at a furniture factory in Newark (owned by ex-Communist friends of my parents) and raised the rest in the West End Bar dealing small quantities of wonderful opiated hashish that a GI had smuggled back from Vietnam. Quite cautious, I sold only to people I knew, but I knew a lot of people. By telling my customers they were actually making a contribution to the cause, financing my Cuba trip, I was able to move slivers weighing only a few grams for ten dollars, a high price at the time. Business was quite brisk. That was my introduction to political fund-raising.

Our group of twenty SDSers left Chicago for Cuba on January 31, 1968, the same day that the historic Tet Offensive began in Vietnam. Forces from the combined National Liberation Front of South Vietnam (NLF) and the North Vietnamese army mounted simultaneous attacks that day in 134 cities and towns, including every provincial capital and major city. American television showed footage of the U.S. embassy in Saigon, the symbol of our occupation of South Vietnam, being held for six and a half hours by NLF soldiers. The American people were taken completely by surprise at this news, because all they had heard for years were Pentagon and White House lies that victory was in sight.

The Cubans we met during our three weeks there spoke incessantly of Vietnam and hoped the Tet Offensive would drive the Americans out. Cuba was the United States turned on its head: On La Rampa, one of Havana's main streets, there was a huge neon-outlined map of Vietnam that gave the latest tally of American planes shot down. Posters exhorted people to work harder, *como en Vietnam* (like in Vietnam), comparing production to the battlefield against U.S. imperialism. I felt as if I had gone through the looking glass. All my beliefs that at home seemed so radical, outrageous, and even illicit were the norm here. When our tour bus pulled in to a small town, people greeted us with the latest news of battles in Vietnam, as if it were the World Series. Encounters like this mesmerized and energized me. I was in heaven.

In Havana our group visited with the Vietnamese delegation to Cuba, which included soldiers and Communist Party members as well as diplomats. I was quite surprised at how warmly we were greeted, given that we were Americans and that our government was murdering Vietnamese as we spoke.

The Vietnamese ambassador began by stating that his people desired only friendship with Americans and that they recognized the great difference between the U.S. government and the American people. Toward the end of the hours-long session, when I got a chance to speak, I said I wished the ambassador's opening statement were true, but the majority of Americans supported the government and the war. (I didn't yet realize that public opinion was flipping in response to Tet.) Our Vietnamese host smiled and patiently answered me, "This will be a very long war. It has already lasted, for us, more than twenty years. We can hold out much longer. We're not going anywhere. Eventually the American people will tire of the war, and will turn against it. Then the war will end."

Of course he was proved right.

Another Vietnamese delegate told us he had been posted by the Communist Party away from his home village and hadn't seen his wife and family for five years. A third had fought in the south for seven years, living for long periods of time—whole years—in tunnels, coming out only at night to attack the Americans. Meeting people like this, who were willing to sacrifice so much for their independence, I could see that the United States didn't have a chance of victory. And I could no longer consider the Vietnamese as a faceless enemy; they were my heroes.

At the end of the meeting, we were each given souvenir rings made of extremely lightweight titanium. The number 2017 was stamped inside to indicate that each ring had been made from debris from the 2017th American plane shot down in Vietnam. I wore mine proudly for years afterward.

Cuba seemed to me a society inspired by a new morality. Cubans we met appeared to be working selflessly, not for their own gain but to improve the whole society. Young people were in charge—the director of a school for teachers located on a mountaintop was twenty-three years old; his counterpart at a rural clinic was in her late teens. I wandered around Havana in a euphoric haze, entranced by the thought that everything I saw—the buildings and the shops and the parks and the doorknobs—all were now produced and owned by the state, by the Cuban people themselves, not the capitalists and landowners. Students, workers, and artists we met spoke to us of working not for material advantage but for "moral incentives," meaning building their country for all the people. They claimed they were creating the new socialist man, a notion first advanced by Comandante Ernesto "Che" Guevara. Cuba's socialism was the moral and political wave of the future. I was stoned on socialism!

Everywhere we went—in store windows, in people's homes, in public buildings—we saw portraits of Che, the young Argentine doctor turned guerrilla who had accompanied Fidel Castro to the Sierra Maestra in 1956 and then commanded the second and deciding front in the revolutionary war to overthrow the dictator Fulgencio Batista. Rounding a bend on a mountain road, our group looked down into the valley below, amazed at the sight of a giant Che profile, at least an acre in area, made out of white rocks against the deep brown earth of the newly plowed field, his beard and beret instantly recognizable.

Che had been murdered in Bolivia just a few months before, in October 1967. Fidel had proclaimed 1968 "The Year of the Heroic Guerrilla" to honor his friend's memory—and his strategy. A few months before his death, Che had written a famous message of undying militancy to the global liberation struggle:

> Our every action is a battle cry against imperialism and a battle hymn for the people's unity against the great enemy of mankind: the United States of America. Wherever death may surprise us, let it be welcome, provided that this, our battle cry, may have reached some receptive ear, and another hand may be extended to wield our weapons and other men be ready to intone the funeral dirge with the staccato singing of the machine guns and new battle cries of war and victory.

Staring at Che's famous profile on the valley floor below, I felt the power of the entire Latin American revolution. The campesinos who had honored him with their labor were the very people who would rise up and seize power. Comandante Che was the symbol of all those who would eventually build a new world. In death he would live: *¡Viva la Revolución!* Victory or Death!

Che—revolutionary martyr and saint—touched my twenty-year-old soul. He had written, "The duty of every revolutionary is to make the revolution," meaning that it's not enough to talk about revolution, you have to take action. Like a Christian seeking to emulate the life of Christ, I passionately wanted to be a revolutionary like Che, no matter what the cost. After only three weeks in Cuba, I was burning to get back to New York and make my own contribution. These Vietnamese and Cubans were fighting on their fronts; I knew that I belonged "in the belly of the monster," to quote a phrase by José Martí, the father of Cuban independence.

Another of Martí's maxims, also quoted by Che, rang in my head like a mantra: "Now is the time of the furnaces, and only light should be seen."

When I returned to New York in early March, I discovered that the American political environment had undergone an astounding change. Following the Tet Offensive, public-opinion polls now showed that fewer than 24 percent of the voters approved of LBJ's conduct of the war. Also, Eugene McCarthy's "dump Johnson" candidacy was showing signs of strength. The little-known poet-senator had almost beaten LBJ in the New Hampshire primary. Robert Kennedy announced his candidacy a few days later, and he also attacked Johnson's now-vulnerable war policy.

On my first day back, I found myself in an SDS general-assembly meeting that was debating whether SDS should support McCarthy or Kennedy or neither. Almost every SDSer spoke of the "McKennedy" peace candidacies as a trick to "co-opt" the energy of radical young people into working for meaningless reforms put forward by hypocritical liberals. McCarthy's "Clean for Gene" slogan implied to us more than a practical call to cut our hair, put on straight clothes, and campaign for a peace candidate: It was the liberal enemy trying to destroy our movement. Liberals, including Robert Kennedy, his martyred brother John, and LBJ had given us Vietnam in the first place.

Our goal was a much more fundamental change, not just ending the war but ending the capitalist system that had caused the war. I didn't support any candidate in 1968, nor did most other SDS members. Electoral politics was beneath our concern.

Except where it came to SDS's internal politics. The Columbia SDS was holding its annual election of officers in March, soon after I returned. I feared that a new leadership would be elected that would continue the tactical conservatism of the past two years. SDS had held peaceful demonstrations, petition campaigns, polite debates, and educational forums concerning the university's involvement with the war and also its expansion into the Harlem community. It seemed to me that now was the time for action. The huge confrontational demonstrations at the Pentagon and in Oakland in the fall had moved us "from protest to resistance," as the slogan coming out of the SDS National Office in Chicago said.

I helped form a caucus composed predominantly of the younger people in the chapter—the freshmen and sophomores—plus JJ and a few other juniors. We quickly became known as "The Action Faction." I ran for chapter chairman. In the end, in order to defeat a rival Progressive Labor Party slate, the Maoist bunch attempting to infiltrate Columbia SDS to recruit new members, we joined with our friendly opposition, led by the two Teds, Kaptchuk and Gold. The Praxis Axis, as the old-timers were now known, put up Nick Freudenberg for vice chairman, a soft-spoken and smart sophomore, and our combined slate easily defeated the candidates from PL.

As chairman I attended all the various subcommittee meetings, in addition to carrying out my formal duties of chairing the Steering Committee and general-membership meetings. I must have spent at least four or five hours a day in meetings or otherwise organizing for SDS. My intense activity was not unique: That spring our SDS chapter had twenty to thirty members who were devoting all our energy to the work. In addition, approximately seventy-five people gave several hours each week, and we had an outer circle of several hundred dependable supporters and sympathizers. A lot of energy was going into radical organizing.

Despite the emergence of the Action Faction, most of my SDS comrades were still quite cautious. The SDS Draft Committee decided, against my suggestion, that we would greet a speech by the head of the Selective Service System for New York City, a Colonel Akst, with "probing questions." I was outraged: SDS intended to sit politely and listen to a speech by one of the head procurers for the war—a war criminal—then ask some embarrassing questions. What wimps! We needed to confront these people with more than words if we were ever going to build a serious movement.

I conferred with Tom Hurwitz, a tall, lanky junior, the son of a left-wing filmmaker, and several other, younger, "Action Factionists." We came up with a tentative plan. Tom then talked with some friends in Up Against the Wall, Motherfucker! the anarchist street gang from the Lower East Side that had just become an SDS chapter, and they promised to help out. We still lacked the lead player for our little skit.

By chance Tom encountered a long-haired hippie named Lincoln Pain who was passing through town from Berkeley, California, the New Left and countercultural epicenter. Lincoln talked nonstop about revolutionary this and militant that. Tom told me on the phone, "I think I got our guy," and I immediately hopped a subway down to his apartment in the Village, where we finished planning the caper.

On a brisk March afternoon, Colonel Akst, stuffed into his uniform, pudgy red cheeks shining beneath his proud cap, stood up at the podium of Earl Hall, the campus religious center, to deliver his patriotic message. I was sitting right in front of him, holding a big white bakery box on my lap, too excited to be scared. Just as he was beginning his talk, a commotion broke out in the back of the hall.

The several hundred people who had packed the enormous domed room turned to see what was going on. A fife-and-drum corps, dressed in mock–Revolutionary War uniforms and armed with toy guns, was playing a ragged rendition of "Yankee Doodle." Turning back to the front, as if with one head, the crowd let out an enormous collective gasp: The distinguished colonel's face was plastered from cap to chin with lemon meringue pie!

By then Lincoln and I were at the fire door, which Jeff Sokolow, a sophomore Action Factionist whom I had recruited minutes before, held open so the guerrilla pie thrower could make his escape. Unfortunately, Lincoln put on his bandanna *after* we left Earl Hall, so there we were, three guys running down Broadway, one of them a criminal pie thrower wearing a bandanna over his face. At my girlfriend Sue's apartment on West 115th Street, I fumbled for several seconds with the keys for the three dead bolts, and when we finally got in, Lincoln went directly to a closet and jumped in. No one, of course, was chasing us.

The guerrilla-theater action worked perfectly, with the exception of the fact that I had forgotten to have a photographer there to record the event, something I've always regretted. But that oversight was partially redeemed by a university administrator with the appropriate name of McGoey (pronounced McGooey), whose memorable quote appeared in the *Spectator* the next morning. If the pie was thrown by a person not associated with the university, he said, "we'll throw the book at him." I told the paper I'd heard that the "New York Knicker-boppers" had done the dirty deed.

The only people on campus—besides the administration—who didn't like the action were the old leadership of SDS, the Praxis Axis. They thought it was "unserious and terroristic"; I thought they were nuts.

The pie affair didn't seem to hurt our campaign against Columbia's participation in the war: By March 27, a week after the notorious incident, SDS had fifteen hundred names on a petition calling for the severing of the university's ties with the Pentagon think-tank, the Institute for Defense Analyses (IDA). On that day I led a crowd of 150 SDS members and others in noisily presenting the petition to a university vice president, President Grayson Kirk being "unavailable." For an hour we roamed through Low Library, the central administration building, talking with workers there about the university and the war and effectively disrupting work.

A year before, two SDS researchers, sophomore Bob Feldman and graduate student Michael Klare, working in the library stacks, had found a reference to Columbia's membership in IDA, an obscure twelve-university consortium virtually unknown to students and faculty. IDA eventually provided one of the ignition points for the explosion about to occur.

In the age of technological warfare, the Pentagon had set up semi-autonomous "think tanks" such as IDA as a means of channeling scientific talent toward developing weapons systems. Columbia's intimate connection to IDA had facilitated its professors' technical research for the Pentagon's Weapons Systems Evaluation Group. In the 1950s, IDA research had concentrated on improvements in thermonuclear weapons and ballistic-missile delivery systems, reflecting the Cold War arms race against the Soviet Union. With the onset of the Vietnam War, IDA was given responsibility for the development of such techniques and weapons as the use of chemical herbicides to destroy the insurgents' jungle cover—the horrible "defoliation" using highly toxic Agent Orange, and the use of airpower for counterinsurgency—"carpet bombing." IDA's elite Project Jason, of which Columbia professors were members, produced reports with titles such as "Interdiction of Trucks from the Air at Night" and "Data Related to Viet Cong/North Vietnamese Army Logistics and Manpower." There was even an IDA report on the suppression of ghetto rebellions. Defense research apparently included domestic counterinsurgency against a portion of the American population.

Along with IDA, the university also had contracts with governmental agencies for other military- and foreign-affairs-related research worth $58 million, which was a staggering 46 percent of the university's budget at the time. The CIA and the State Department were major funders of the School of International Affairs and the various area institutes like the Russian and the East Asian institutes. Columbia had even founded a brand-new Southeast Asian Institute to help the government plan strategic war policy.

IDA thus became the shorthand symbol for Columbia's huge network of complicity with the war. For months the administration would not even answer our inquiries, letters, and petitions. The Graduate Faculties dean told an audience concerned about IDA that "Columbia has no institutional connection with the Institute for Defense Analyses." When confronted with the facts concerning the institution's lucrative ties to IDA, another university administrator explained, "These things are not in the purview of faculty and students." And another vice provost said, "A university is definitely not a democratic institution. . . .Whether students vote yes or no on an issue is like telling me they like strawberries." (This remark would give rise within the year to the title of the delightful bestseller *The Strawberry Statement,* by James Simon Kunen, a Columbia sophomore.)

The vice president's response inflamed our anger almost as much as the war research itself did. Who is the university? we demanded to know. Is it only the administrators and trustees, or is this not a "community of scholars," which includes faculty and students? If it's *not* a community of scholars, *why* isn't it?

By demonstrating against IDA in Low Library on March 27, we openly defied a ban on "indoor demonstrations." Five members of the SDS Steering Committee, including myself and Ted Gold, plus the Columbia chairman of the antidraft organization The Resistance were called in for disciplinary action. We refused to appear, demanding a public hearing on our case before the whole university. Our strategy was to use the IDA 6, as we called ourselves, to publicize the university's nonneutrality.

In response to SDS's petition demanding the university's disassociation from IDA, Kirk announced a new "study group" on the matter. Study groups were an old tactic used many times by the university to kill a controversy. On April 1 the trustees of Columbia and the directors of IDA announced what SDS called the "April Fool's Plan": They claimed that the university was now no longer "institutionally" associated with IDA, though Grayson Kirk still sat on IDA's executive committee, which approved all projects. Obviously, they were trying to give the appearance of restructuring while substantively changing nothing. We were surprised that they underestimated our ability to see through this transparent sham, but at the same time we were delighted to note how defensive they had become about Columbia's business as usual.

The evening of Sunday, March 31, 1968, I was riding in a car with some other Columbia SDSers, returning from my first national SDS meeting, in Lexington, Kentucky. The debates over SDS's future program and direction had been chaotic, with my anarchist friends from the Lower East Side, the Motherfuckers, disrupting every speech they didn't like with the cry, "That's bullshit, and you know it!" But I was still euphoric from hanging out with SDS members I had not known from other chapters around the country.

We were outside Cincinnati when the radio blared LBJ's historic message, "I shall not seek, and I will not accept, the nomination of my party for another term as your president." We stopped the car, whooping and hollering, hugging each other and dancing on the side of the road. Victory! We had done it! The antiwar movement had forced a sitting president to abdicate and seek peace. Johnson had agreed to call a limited bombing halt and to begin talks with the Vietnamese.

Over the next few years, as the war dragged on, our momentary euphoria would be replaced by pessimism and weariness. Yet years later, when the full story came out, I learned that our immediate reaction was closer to the truth. The release of the Pentagon Papers in 1971, plus the later testimony of administration insiders, revealed that LBJ's senior advisers, retired bigwigs called the "Wise Men," and his hawkish new secretary of defense, Clark Clifford, had advised him in March 1968 not to escalate the war. They admitted that *the war could not be won.* The single greatest limiting factor, the Wise Men observed, was the American people's inability to support the ever-increasing military effort Johnson and the generals wanted. The United States already had a half million troops in Vietnam, and Congress had turned down a surtax to finance the war, believing that it would be politically untenable.

After March 31, 1968, the American problem was how to get out of Vietnam, not how to win; the war was already lost, and the war managers knew it. Even so, they dragged on the carnage for seven more years, with approximately half the American casualties, thirty thousand names on the wall in Washington, occurring *after* the Tet Offensive. (There's no way to know how many Vietnamese they murdered in this time.) This taught me never to underestimate the willingness of cynical old white men in Washington to sacrifice other people's children.

Four days later, on April 4, Martin Luther King Jr. was assassinated in Memphis, Tennessee. He was there helping sanitation workers who were on strike to force the city to recognize their union. Black garbage collectors, supervised by whites in a southern city, had endured abysmal conditions and pay. In King's view the question of poverty was as much a civil-rights issue as the right to vote or eat at a lunch counter.

In response to the assassination, black ghettos in almost a hundred cities and towns around the country went up in flame, including Harlem, right beside Columbia.

Standing across from Columbia president Grayson Kirk's mansion, at the corner of Morningside Drive and West 116th Street, JJ and I looked out over Harlem as dusk fell that unseasonably warm early April night. Below, we could see the flames of dozens of fires, dark smoke clouds trailing upward. In the distance we heard the shrieks of fire and ambulance sirens and, rising above that, the roar of a sound I've never heard before or since: the wailing of the hundreds of thousands of people of Harlem, the capital of black America.

JJ and I looked at each other, and both of us said the same words at the same time: "Let's go!" We tore down the broken steps that descended through untended, garbage-strewn Morningside Park, the no-man's-land between the Heights and Harlem. Wandering around Harlem most of the night, we watched people loot TVs and stereos from appliance stores, face down lines of beleaguered cops, set up improvised barricades of garbage cans to block the fire engines. "Burn, baby, burn!" seemed to be the message. Along the way, many concerned black people warned us that it wasn't safe for whites on the street, yet we were never once threatened. Strangely, we saw an anarchic joy of release in some people's faces, mixed with the rage. Others were standing around crying, while still others were "dancing in the streets," as Martha & the Vandellas sang.

After a few hours, JJ and I got separated, and I wandered around alone, invisible, drinking it all in. Malcolm X, assassinated more than three years before, was alive that night on the streets of Harlem. I saw him.

For the next few days, the atmosphere at Columbia mirrored the intensity of the world outside. Every conversation had to do with the persistence of racism in society and in our own lives. This was when many people at Columbia first became aware of the fact that the university planned to build an eleven-story gym in Morningside Park.

Back in February, when I was in Cuba, Columbia SDS's Neighborhood Committee had become involved with a group of Morningside Heights and West Harlem residents who had for many years been fighting the building of the gym. The building would have a separate entrance at the bottom for black Harlem residents, who had 15 percent of the space in the facility allocated to them. Under the slogan "Gym Crow Must Go!" opponents had greeted the first two weeks of construction of the gym in February with three demonstrations. Twenty-six people—both students and community members—were arrested. But few white people at Columbia had paid attention to the issue before the King assassination put institutional racism right into their faces, where they could no longer ignore it.

At that same moment, the hundred or so black students on campus, fewer than 1 percent of the university's student body, were reconsidering their relationship to this white institution. Within a few days of Dr. King's murder, they elected Cicero Wilson, a tough, barrel-chested, young black student who was said to be "street," to the position of chairman of the Student Afro-American Society. Along with a core group of politically savvy student-organizers, he promised to reverse the direction of what had up to then been an apolitical if not slightly conservative social organization. Blacks had just begun to be admitted to Columbia College and Barnard, and their numbers remained small—fifteen to thirty-five per class. Reflecting the segregated nature of the school, security guards constantly stopped the black students on campus to check if they actually belonged there. By taking up the issue of the gym, the black students announced to the whole campus that they were representing the black community surrounding Columbia.

Columbia had scheduled an official memorial for Dr. King; it was the least a liberal white institution could do. This struck us in SDS as pure hypocrisy, because of what we knew about Columbia's racist treatment of the community and its black and Latino service workers. So we held an emergency meeting of our chapter's Steering Committee, which I chaired, to discuss the upcoming memorial and to plan an impromptu action for the next day.

On April 9, about a thousand people filled St. Paul's Chapel, the massive classically ornate university chapel beside Low Library, to attend Columbia's memorial service for Dr. King. President Kirk was present. As the audience entered, SDS members handed them leaflets detailing the university's mistreatment of black and Latino employees, its eviction of over ten thousand mostly nonwhite residents from

Morningside Heights buildings, and its theft of land from a public park to build a segregated gym.

I was sitting near the front of the audience, dressed in a jacket and tie. Just as Vice President David Truman was about to deliver his eulogy, I walked up to the podium, trembling inside with fear, and stepped in front of him. I grabbed the microphone and forced myself to begin speaking in a quiet, steady voice, "Dr. Truman and President Kirk are committing a moral outrage against the memory of Dr. King." The PA system went dead.

I continued anyway. "How can the leaders of the university eulogize a man who died while trying to unionize sanitation workers when they have, for years, fought the unionization of the university's own black and Puerto Rican workers?" I asked. "How can these people praise a man who fought for human dignity when they steal land from the people of Harlem? And how can these administrators praise a man who preached nonviolent civil disobedience while disciplining its own students for peaceful protest?"

The audience sat in stunned silence. "Dr. Truman and President Kirk are committing a moral outrage against the memory of Dr. King," I repeated. "We will therefore protest this obscenity."

I stepped away from the lectern and walked down the center aisle and out the main door. I was followed by about forty people, including not only SDS members but several black students and an elderly black couple from the community.

We stood around outside and talked among ourselves as the service continued in the chapel. The general feeling was that we had done what needed to be done; people congratulated me on my comments, which I had worried weren't heard because the microphone was cut off. Several people confirmed that the university had for many years blocked black and Latino cafeteria workers from organizing, using classic union-busting tactics of intimidation, buying off some workers and firing the leaders. The last unionization effort that Columbia beat back was in 1964, despite a cafeteria boycott organized by sympathetic students in Columbia CORE, many of whom later joined SDS.

The disruption at the King memorial was front-page news in the next morning's *Spectator.* The campus was buzzing about it; many people viewed the protest and walkout sympathetically, although in subsequent days the paper was also filled with letters critical of me and SDS. The university chaplain, the Reverend John D. Cannon, publicly defended my right to speak, an act that helped legitimize our protest and also thoroughly enraged the administration.

A week later I read in the *Spectator* that President Kirk had made a speech at the University of Virginia about the war in Vietnam and its effect on U.S. college campuses. On April 12, three days after the King walkout, he said, "Our young people, in disturbing numbers, appear to reject all forms of authority, from whatever source derived, and they have taken refuge in a turbulent and inchoate nihilism whose sole objectives are destructive."

Such a dummy, I thought. *This guy doesn't understand anything at all.* He did seem to have an inkling that Vietnam had something to do with this generation gap, because in the speech he called for the United States "to extricate itself as quickly as possible from its current involvement in Vietnam." Kirk was such a ruling-class liberal, though, that he couldn't leave it at that. He referred to the war as "a well-meant but essentially fruitless effort" made by "sincere, honorable, and patriotic men who do not deserve the calumny to which they have been subjected."

I took it on myself to set Grayson Kirk straight.

On April 22, Columbia SDS published a four-page newspaper entitled *Up Against the Wall!* We circulated it throughout the campus in order to drum up support for the showdown with the administration that we knew was coming the next day. We had planned another indoor demonstration in support of the IDA 6. It seemed clear to us that the administration was trying to kill the antiwar movement on campus by picking off the leaders, so we would again violate their rule en masse.

Up Against the Wall! contained several articles on the war and the university's complicity with it, the administration's attempt to repress SDS on campus through the use of threats of discipline, and a defense of the right of civil disobedience. Featured on the front page was the following open letter from me to the president of Columbia University. It was entitled "Reply to Uncle Grayson."

Dear Grayson,

Your charge of nihilism is indeed ominous: for if it were true, our nihilism would bring the whole civilized world, from Columbia to Rockefeller Center *[built on Columbia-owned land]* down upon all our heads. Though it is not true, your charge does represent something: you call it the generation gap. I see it as a real conflict between those who run things now—you, Grayson Kirk—and those who feel oppressed by and disgusted with the society you rule—we, the young people.

You might want to know what is wrong with this society, since, after all, you live in a very tight, self-created dream world. We can point to the war in Vietnam as an example of the unimaginable wars of aggression you are prepared to fight to maintain your control over your empire (now you've been beaten by the Vietnamese, so you call for a tactical retreat). We can point to your using us as cannon fodder to fight your war. We can point out your mansion window to the ghetto below [that] you helped to create through racist university expansion policies, through your unfair labor policies, through your city government and your police. We can point to this university, your university, which trains us to be lawyers and engineers and managers for your IBM, your Socony-Mobil, your IDA, your Con Edison *[Kirk sat on these corporate boards]* (or else to be scholars and teachers in more universities like this one). We can point, in short, to our own meaningless studies, our identity crises, and our repulsion with being cogs in your corporate machines as a product of and reaction to your basically sick society.

Your cry of "nihilism" represents your inability to understand our positive values. If you were ever to go into a freshman CC class you would see that we are seeking a rational basis for society. We do have a vision of the way things could be: how the tremendous resources of our economy could be used to eliminate want, how people in other countries could be free from your domination, how a university could produce knowledge for progress, not waste consumption and destruction (IDA), how men could be free to keep what they produce to enjoy peaceful lives, to create. These are positive values, but since they mean the destruction of your order, you call them "nihilism." In the movement we are beginning to call this movement "socialism." It is a fine and honorable name, one which implies absolute opposition to your corporate capitalism and your government; it will soon be caught up by other young people who want to exert control over their own lives and their society.

You are quite right in feeling that the situation is "potentially dangerous." For if we win, we will take control of your world, your corporation, your university and attempt to mold a world in which we and other people can live as human beings. Your power is directly threatened, since we will have to destroy that power before we take over. We begin by fighting you about your support of the war in Vietnam and American imperialism—IDA and the School of International Affairs. We will fight you about your control of black people in Morningside Heights, Harlem and the campus itself. And we will fight you about the type of miseducation you are trying to channel us through. We will have to destroy at times, even violently, in order to end your power and your system—but that is a far cry from nihilism.

Grayson, I doubt if you will understand any of this, since your fantasies have shut out the world as it really is from your thinking. Vice President Truman says the society is basically sound; you say the war in Vietnam was a well-intentioned accident. We, the young people, whom you so rightly fear, say that the society is sick and you and your capitalism are the sickness.

You call for order and respect for authority; we call for justice, freedom, and socialism.

There is only one thing left to say. It may sound nihilistic to you, since it is the opening shot in a war of liberation. I'll use the words of LeRoi Jones, whom I'm sure you don't like a whole lot: "Up against the wall, motherfucker, this is a stick-up."

Yours for freedom,
Mark

4

Columbia Liberated

Well before the advertised time of noon, the crowd started forming at the Sundial, a circular raised platform at the center of campus. Many people came anticipating a showdown with the administration over the issues of the gym in Morningside Park, the Institute for Defense Analyses, and the disciplining of the IDA 6. Others were there because they expected a confrontation with the right-wing students, the jocks, and wanted to make a stand with SDS.

The previous Friday an anonymous anti-SDS leaflet had been circulated with the question "Tired?" written twenty-one times across the top of the page. At the bottom it asked, "Can democracy survive at Columbia University? Will Mark Rudd be our next dean? Be there on the 23rd—prepared." Over the weekend we SDSers had copied the leaflet in large quantities, inserting our own comments pointing out the threats in the margins, and circulated it as widely as possible. Our countertactic seemed to have worked: We had the numbers by far.

The right-wing Students for a Free Campus were picketing and standing around at the top of the steps leading to Low Library, presumably guarding the administration offices. But there were only 150 of them. They carried a small number of signs with messages like SEND RUDD BACK TO CUBA! and ORDER IS PEACE. On the next level of the steps, below them, by the bronze statue of Alma Mater, several hundred faculty, students, and other curious spectators milled around, waiting for the action to begin. At the lowest level, College Walk, the part of West 116th Street closed off to traffic, about five hundred people gathered around the Sundial to hear the speeches and then march to Low, an action the SDS Steering Committee had planned the night before. At that meeting we'd had no idea how many people would show up.

Ted Gold and Nick Freudenberg from SDS spoke first, passionately attacking IDA research as Columbia's complicity with the war in Vietnam and the gym in Morningside Park as a symbol of the university's racist stance toward the neighboring Harlem community. The crowd was attentive, expectant.

Around midnight the night before, I had met with Cicero Wilson, the new leader of the Student Afro-American Society (SAS) at my kitchen table. Reflecting the state of segregation at Columbia, I hadn't known him at all, but a freshman in SDS, Paul Berman, played in a soul-music band with him and had brought the two of us together. Cicero said that he was concerned about the right of students to protest and also that SAS would have support from Harlem ministers. He agreed to address the rally. This was a first for both SAS and SDS.

After Ted and Nick spoke, Cicero got up on the Sundial and attacked Columbia for its racist policies but then lambasted white students for not doing anything. This seemed to confuse the mostly sympathetic audience judging by the fact that some in the crowd answered him back.

I was supposed to give the final talk, a rabble-rouser to fire up the crowd before we marched into Low Library, an intentional mass violation of the rule against indoor demonstrations. But just as the rally began, while Nick was speaking, an assistant dean handed me a letter from Vice President Truman proposing a meeting with us in the university's McMillin Theatre. His intention was clear: He wanted to chill us out by getting us into a controlled environment. I suggested instead that the session be an open hearing, with the students determining what discipline should be meted out to the IDA 6. The assistant dean told me this was unacceptable.

Traditionally, all SDS decisions were made by group meetings or, if possible, by consensus of those present during an action. Leaders were there to facilitate debate, clarify issues, and "let the people decide," as the SDS slogan said.

Standing on the Sundial, I read Truman's letter to the crowd—with a few sarcastic additions, ending by saying, "He gives us this alternative because he is a very *li-ber-al* man. After we've gone to the son of a bitch a million times and he hasn't responded to us, now he wants to meet with us in McMillin." The crowd laughed.

I really wasn't sure what to do, so I tried to buy time by weighing the alternatives out loud. "We could have a demonstration inside McMillin, with chanting and picketing. . . . But if we go to McMillin, we will just talk and go through a lot of bullshit." Then I informed the crowd that the doors to Low had been locked, which I learned from an SDS runner who signaled me from the top of the stairs. Just as I was about to propose going to Hamilton Hall, where the Columbia College dean's office was located, Tom Hurwitz of SDS's Action Faction, jumped up on the Sundial and shouted, "Did we come here to talk or did we come here to go to Low?"

People were yelling, "To Low, on to Low!" and the crowd began flowing up the steps. I and the other five of the IDA 6 ran to the front of the crowd—we were supposed to be the leaders, after all—and linked arms as we chanted, "IDA MUST GO! IDA MUST GO!" Here I was, at the head of a demonstration about to burst into a locked building, or else run headlong into a mob of pissed-off right-wing jocks, and I had only the vaguest idea of what we were doing. I knew we wanted to disrupt Columbia's normal functioning in order to provoke further confrontation with the administration. This sort of disruption had recently been successful at other schools, like the University of Wisconsin and Howard University in D.C., and it had caused many more people to rally to the radical side. But what should we do with the locked doors? Where to go if we couldn't get in? What to do if the jocks attacked us?

When we reached the top of the steps, we found that the doors were indeed locked. About twenty-five SDSers tried to force their way into a side entrance, but campus cops pushed from the inside and we backed off, instinctively fearing violence.

I got up on an overturned trash can, intending, once again to specify the options to the crowd. But before I could say anything, someone yelled, "To the gym site!" From the trash can, I watched three hundred people stream off toward Morningside Park, several blocks from campus. I yelled after them, "Tear down the fucking fence!" then jumped off and followed along, passing through a crowd of seething jocks. On the way to Morningside Park, I encountered Ted Gold and Ted Kaptchuk, the old leaders of the SDS chapter, both of whom were upset. "Your demonstration's out of control," Kaptchuk chastised me.

"I know," I replied. "I have no idea what to do."

Ted Gold was a problem solver. "There's a lot of people still at the Sundial. I'll go and try to rally them."

"Right," I replied. "I'll see what's happening at the gym site."

I had never been to the site before, but all I had to do was follow the running crowd and the noise of police sirens. By the time I got there, the demonstrators had already in fact started tearing down the fence. A cop who was guarding the construction had wrestled one of the students to the ground, and the crowd had jumped the cop, beating and kicking him. When I arrived, the scene was at a standoff, with one student in handcuffs—a guy named Fred Wilson, who wasn't a member of SDS—sirens wailing, and police reinforcements racing down the hill. People were barraging me with demands that I quiet the crowd and stop the violence.

I climbed up on a mound of construction dirt and began to speak, but the bulldozers drowned out my voice. I turned to a police captain, one of about a dozen cops surrounding me, and demanded, "Get that stuff turned off so I can talk!" He signaled to a cop closer to the machines, who passed the sign along. The diesel engines went quiet.

I addressed an assistant dean standing on the periphery of the crowd. "You have fifteen minutes to get that guy unbusted," I threatened. "We won't leave until that guy comes back. If not, we'll shut the site down. Get up to Low Library and see Truman now."

The dean didn't move, and neither did I on my little hill. A debate broke out among the demonstrators about what to do. I suggested we organize a strike for the next day. This was shouted down as being unrealistic: A strike couldn't be organized on such short notice. Finally someone suggested going back to the Sundial, and so, for want of a better plan, and because we were now surrounded by dozens of police, the crowd started back toward campus.

I was totally dejected by now. The day had been an utter disaster. The original crowd of five hundred demonstrators had gone from being an organized body, with a purpose and leadership, to a loose mob of stragglers spread out between the campus and the park. One of our members had been arrested, and we were powerless to do anything about it. SDS's cumbersome democratic decision making had ensured that no decisions could be made. We were acting without a plan, and I, the big leader, had no notion of what to do.

As I turned the corner from Morningside Drive onto 116th Street, I could see a large crowd of students marching in our direction. Ted Gold and Ted Kaptchuk had rallied those left back on campus, three hundred strong, to march to the gym site. Those of us straggling back from the site linked up with them, and together we marched back to campus. Perhaps all was not lost yet.

Everybody and his uncle was grabbing my coat and giving me advice about what to do: "Mark, let's take over Kent Hall"—the university computer center. "Mark, we should call a strike for tomorrow and go home and get on the telephone and call everyone." "Mark, you should act more aggressively." "Mark, you should defuse the anger in the crowd." Two nice things about Columbia students: Every one of them has an opinion, and none of them were shy about sharing it. No passive followers need apply. I sent out runners to find possible sites for an indoor demonstration. That would at least accomplish our original objective of violating the university's ban on indoor demonstrations and would also give us a place to reorganize ourselves.

Back at the Sundial, I reassumed the podium. I thought it was best to be honest. "We're learning to criticize and learn from our mistakes," I began. That was radicals' code for, "We messed up."

"The way I see the situation, we've got about four hundred, five hundred people who'll do anything now," I said. The crowd cheered. "On the other hand, I don't know if we've got four hundred or five hundred people who'll do anything tomorrow—but I think you will." I still didn't know what the conclusion would be, so I thought out loud. "I don't think four or five hundred people can close down this university." I *still* hadn't a clue, so I yielded the Sundial to Cicero Wilson, chairman of the Student Afro-American Society.

Wilson wasn't in much better shape than I. At first he seemed to be praising the whites for our strong action at the gym site; then he vaguely implied that the black students would now take over the demonstration. "SDS can stand on the side and support us," he said, "but the black students and the Harlem community will be the ones in the vanguard." At the end *he* wasn't able to propose a direction either.

Bill Sales, a brilliant and eloquent graduate student in the School of International Affairs and a leader of SAS, jumped onto the Sundial. "Okay, I want you to check something out," he began. "I thought up until this stage of the game that white people weren't ready. But I saw something today that suggests that maybe this is not true. Maybe you are ready. Because when the deal hit the fan, you were there, you were with me.

"If you're talking about revolution, if you're talking about identifying with the Vietnamese struggle . . .you don't need to go marching downtown. There's one oppressor—in the White House, in Low Library, in Albany, New York. To strike a blow at the gym, you strike a blow for the Vietnamese people." The crowd cheered. "You strike a blow at the gym and you strike a blow against the assassin of Dr. Martin Luther King Jr. You strike a blow at Low Library and you strike a blow for the freedom fighters in Angola, Mozambique, Portuguese Guinea, and Zimbabwe, South Africa." Everyone was cheering: We loved it.

He ended by advocating "superior organization and superior commitment" so that we wouldn't act like an "incoherent mob." More loud cheers.

It was my turn again. "We don't have an incoherent mob, it just looks that way. I'll tell you what we want to do. We want to win some demands about IDA, we want IDA to go. We want the people under discipline to get off of discipline. We want this guy who got busted today [I couldn't remember his name] to get the charges dropped against him. We want them to *stop* the fucking gym over there. So I think there's really only one thing we have to do, and we're all together here, we're all ready to go—now! We'll start by holding a hostage."

"Where are we going to get one?" somebody asked.

"There's one part of the administration that's responsible for what happened today—and that's the administration of Columbia College." While Cicero Wilson and Bill Sales were talking, one of the runners had reported to me that Hamilton Hall was wide open.

Someone yelled out, "Seize Hamilton!" I shouted, "Hamilton Hall is right over there. Let's go!" The crowd started off, four hundred fired-up people chanting with one thundering voice, "IDA must go! IDA must go!"

Actually, when I said we should hold a hostage that first time, I meant a building. That is, to occupy a building until our demands were met. Sit-ins were time-honored tactics of the labor and civil-rights movements, recently used by students at other campuses, and that was the model I had in mind. But when Acting Dean Henry Coleman walked up to me in the angry, chanting crowd in the tightly packed lobby of Hamilton Hall, I saw a chance for a stronger statement. After all, this was war—or something very close to it.

I stood in the middle of a mob of hundreds of very enraged students, face-to-face with Dean Coleman, a tall, crew-cut ex–football coach, who cradled his pipe in his hand trying to look calm.

The crowd quieted. "Now we've got the Man where we want him!" I yelled. "He can't leave unless he gives in to some of our demands."

The demonstrators roared.

"Now, let me tell Dean Coleman why we're here: We're here because of the university's bullshit with IDA. After we demand an end to affiliation in IDA, they keep doing research to kill people in Vietnam and in Harlem. That's one of the reasons we're here. We're here because the university steals land from black people, because we want them to stop building that gym. We're here because the university busts people for political stuff, as it tried to bust six of us, including myself and five other leaders of SDS for leading a demonstration against IDA. We're not going to leave until that demand, no discipline for us, is met."

Sustained applause.

"Another demand is that our brother who was busted today—he got some sort of assault charge—is released, along with all the other people who have been busted for demonstrating over there. So it's clear that we can't leave this place until most of our demands are met."

"Most?" a student yelled from the crowed. "We've got to stay, man."

My turn again. "I just want to ask people, are we disrupting the university's function?"

"YES!"

"Is the university disrupting people all over the world?"

"YES!"

"Are we going to stay here until all our demands are met?

"YES!"

"No deans leave this building?"

"YES!" The crowd started clapping and chanting, "Hell no, we won't go!" the slogan of the draft-resistance movement. After a few moments, the chant died down.

The chief security officer of the university, Proctor William Kahn, a courteous elderly man with whom I'd had many cordial dealings over the previous months, was standing next to Dean Coleman. I pointed to them. "We know that Coleman and Kahn are only lackeys for the Man. We're going to hold them here, but we want Truman and Kirk to come and give in to our demands. We can stay here for a while. If you're hungry, remember that there are a hell of a lot of people suffering because of Columbia University. We've got to put pressure on these guys to change Columbia University."

Pale and shaken, Dean Coleman attempted an answer. "I have no control over the majority of the demands you have made, Mark," he said, "and I have no intention, Mark—I'll make this very clear to you—I have no intention of meeting any demands under a situation such as this." Students began screaming "Fuck you!" and "Bullshit!" at him. He and Proctor Kahn retreated into Coleman's office. Three jocks took up posts in front of the door, arms folded, ostensibly protecting them. Things were happening so fast by that point that I only dimly understood we had passed the point of no return.

Our first task was to form a Steering Committee for the occupation. Quickly I ran around taking names and got suggestions for nine people: three from SDS—Ted Gold, the senior outgoing vice chairman of the chapter; Nick Freudenberg, the new sophomore vice chairman; and myself—three from SAS—Cicero Wilson, Bill Sales, and Ray Brown Jr.—two from the College Citizenship Council—Juan Gonzales and Joel Ziff—and one "unattached liberal," Jon Shils. All men, I should add. Such a male monopoly on leadership would be unthinkable today.

While the Steering Committee met in an upstairs hallway, several hundred students in the lobby sang, listened to speeches, and organized themselves into work crews to handle food, garbage, security, and entertainment. Our job was to come up with a list of demands, a "What We Want." With remarkable unity, we quickly wrote and adopted the following six demands. These same points would come to be ratified time and again throughout the subsequent strike, never varying:

1. That construction of the gymnasium be stopped.
2. That the university cut all connections to IDA.
3. That the ban on indoor demonstrations be rescinded.
4. That criminal charges arising out of protests at the gym site be dropped.
5. That probation for the IDA 6 be rescinded
6. That amnesty be granted for the present protest.

I started to leave to take the list down to the lobby for everyone to discuss, consistent with SDS's standard decision-making procedures. But the three blacks on the committee told me to stay: Our job as leaders, they said, was to make decisions and present them to the others. I assented, but a potential clash in leadership styles was apparent at the outset. Bill Sales and Ray Brown made it quite clear that SDS's hyperdemocratic mode of decision making drove them up the wall.

Our next act was to issue a call for community people, students from other colleges and high schools, and the public at large to join the occupation of Hamilton Hall. We agreed unanimously that the struggle against Columbia should not be limited to students of the university, because much broader issues were involved. Within hours, representatives from SDS at other campuses, the Student Nonviolent Coordinating Committee, the Congress of Racial Equality, and other civil-rights, black-power, and peace organizations were in the building. Food and money donations streamed in, as did individuals from around the city.

Downstairs, in the lobby, a nonstop meeting and rally was in session. Hamilton was still open for students to come and go to classes.

In our cautious, base-building style, SDS did not want to provoke a confrontation with other students. In fact, several dozen jocks were in the lobby with us, giving people dirty looks and even heckling speakers. Five of them had clustered around the door to Dean Coleman's office, taking up posts to protect the two officials inside. Above the door one of the protesters had put a poster of Che Guevara; similarly, giant images of Malcolm X, Stokely Carmichael, and even Lenin peered down at the student rebels from the walls and pillars of the staid old Ivy League building.

At one point that evening, six black men who were not students moved toward Dean Coleman's door. They announced to the jocks that they were there to assume guard duty. The right-wingers refused to move. Suddenly, and without a lot of fuss, the jocks found themselves bounced out of the building; the black phalanx had taken over.

As the protesters began to bed down for the night throughout the building, rumors spread that community people had brought guns into Hamilton. Despite all our talk about revolution, most of the whites, myself included, were scared; I don't know how the black students felt. We were still really middle-class kids, and it suddenly dawned on us we might now be in a different league from the student protest we had begun that afternoon.

The Steering Committee kept meeting until about 2:00 A.M. in an atmosphere of growing discord. The question was what to do in the morning, now that we held a building and the acting dean of the college. The SAS leaders said that the black students, with the support of the community, were ready to block all access to Hamilton and hold out for the demands right there and then. I felt, along with the rest of the whites, that we could not afford to alienate our potential student base by denying them access to classes. Our goal, after all, was to radicalize the student body and draw them into the protest, but we didn't want to create dividing lines with a majority against us. We kept going round and round on the same points; finally we agreed to meet separately with our respective groups, the whites on the seventh floor and the blacks on the third, which they had occupied as an all-black floor.

From 2:00 A.M. until 5:00 A.M., I chaired an open caucus on the seventh floor. I was exhausted and hoarse, but I reported on what had transpired at the Steering Committee: "The blacks want to barricade the building and deny access to the students and faculty. They want to make a stand, now, with the black community here. They're not concerned about having the support of the majority of whites on campus."

In characteristic SDS fashion, a freewheeling discussion ensued that touched on everything from the imperialist ruling class's depredations in Vietnam to the Bolshevik revolution of 1917 and a lot in between. I took a straw poll and found that almost everyone was for keeping open access. But we didn't want a split with the blacks. Reasonably, JJ proposed a politic compromise: We'd keep the administrative offices on the first floor blockaded but allow students and teachers to go to class. Somehow the proposal got lost in all the debate. In the end the SDS caucus decided almost unanimously that we didn't want to block access to Hamilton, but if the blacks went ahead and did so, we would support them. No one, least of all me, wanted to be left out of the action due to our notions of base building. To put it another way, we would not abandon our black brothers.

Just then a runner came in with a message from the black caucus: They had decided to barricade the building. I jumped up and, with Ted Gold and Nick Freudenberg, the other two SDS members of the Steering Committee, raced downstairs to meet with our SAS counterparts. Straight off, Ray Brown, Bill Sales, and Cicero Wilson informed us, "We want to make our stand here. It would be better if you left and took your own building." I was stunned speechless. Despite all our discussion, the break had been forced on us.

The working coalition between SDS and SAS had lasted exactly seventeen hours, from noon to 5:00 A.M.

Probably in part to ease the pain, Bill Sales suggested that by leaving the building we could act as a diversion when the police came, possibly even starting a second front.

Deeply shaken, I slowly climbed the stairs to the caucus room on the top floor. I announced to the people still there what had happened. A few tried to argue with me, but I pointed out that the discussion was over, the split was a fait accompli. People ran through the building waking the sleeping protesters, and then very quietly, dazed and dispirited, we congregated in the lobby. I wanted us to exit together, with some semblance of order.

As we shuffled out the door, one of the freshman SAS members called out to us, "Good luck to you, brothers. We're still together." Tears were in my eyes when I looked back at the black students now beginning to build a barricade on the other side of the door.

As a cold dawn was about to break, three hundred of us white students who had been occupying Hamilton Hall since the previous afternoon streamed out of the ancient classroom building. About half of our number slipped away, probably going home to their warm dorm rooms and apartments. The rest, trailing our books and blankets, silently trudged up the massive stone steps, past the statue of Alma Mater, the female Roman archetype symbolizing the spirit of the university, to the one door we hoped wouldn't be locked, the security entrance in the basement of Low.

The streetlamps glowed in the cold mist. No one talked as we moved toward the security door. It was locked. We had no plan or strategy, other than the single universal thought that we couldn't abandon the blacks back in Hamilton. My buddy JJ and two other SDS guys pried a plank from a wooden bench nearby. They held it up to the large plate-glass window in the door but then hesitated. A second time they held up the plank to the glass but then dropped it down. Finally they took a deep breath in unison and swung the plank together. The window shattered. In the stillness of the morning, the sound of broken glass rang out as it hit the concrete sidewalk. A shudder passed through us—none of us had ever done anything like this before—and then we poured into the building, past the security guards who had given up trying to hold us back. We ran up a flight of steps and found ourselves in the giant black-marble colonnaded rotunda that was the central feature of the monumental neoclassical-style building. I looked up to the coffered dome, fifty or more feet above, amazed at where I was and what we had just done.

We found the entrance to the president's suite. It, too, was locked, but there was a small window in the door. This time there was no hesitation: Someone picked up a wooden signpost marked OFFICE OF THE PRESIDENT OF THE UNIVERSITY, broke the glass, and we were in.

Dozens of us now crowded into the secretaries' antechamber. A door led to the president's own office. We opened it, and . . . Behold! we were in the inner sanctum, a large, book-lined corner room with high ceilings and huge windows looking out over the campus. On one wall was a Rembrandt painting, *Portrait of a Dutch Admiral,* which someone later said was valued at $450,000. The shelves were adorned with priceless Chinese Ming vases.

We were in awe. None of us had ever been in the president's office. Most of us had never seen such a sumptuous room, filled with art, lamps, leather sofas. Then there were the books, seemingly miles of them on built-in shelves. I pulled one down to look at it and saw that the edges of the pages had not yet been cut. Examining others, I found that none showed signs of having been read or even of simple wear. *Hmmm,* I thought, *just as I suspected. This guy's a phony.*

Dr. Grayson Kirk had been president of Columbia University since 1953, when the previous university president, General Dwight D. Eisenhower, supreme commander of Allied troops in World War II, resigned to become president of the United States. No kidding, this was the big leagues. Not only was President Kirk a noted governmental scholar and adviser, he was on corporate and institutional boards: Consolidated Edison of New York, Mobil Oil, IBM, the Council on Foreign Relations (as its president), and, of course, the Institute for Defense Analyses.

Imagine an idealistic Jewish kid growing up in a suburban New

Jersey town, always knowing that the world consisted of two kinds of people: Us and Them, the Jews and the goyim. Crossing the river to the big city and taking a place as a student in a world-class Ivy League institution run by Them, I found at the top, much to my surprise, rather slow-witted, Wizard of Oz–like characters who ran things really badly, violated their own principles, lied, put into effect both pro-war and racist policies. My reaction? In my speeches at rallies, I had taken to referring to President Kirk as "that shithead."

If Kirk and the other people running Columbia had only said, *Oops, we're sorry we lied, but we did it for a good cause. We want to do our part as good American institutional citizens in helping the United States conquer the world. We just didn't think you could handle that*, I might have respected their honesty. But instead they claimed to be upholding a convenient lie called "academic neutrality." Meanwhile, the university was a major military contractor, developing weapons to murder innocent people in Southeast Asia.

The real clincher for me was that the members of the university administration were all liberal men. They were all nominally against the war, but only because the United States was losing, which was the worst reason, in my opinion. Even Grayson Kirk, probably a Republican, was against the war. So was David Truman, the vice president who was the on-campus spokesman and hatchet man for Kirk, the president being too busy fund-raising and sitting on his various corporate boards. Dean Truman, as we called him because he used to be the dean of Columbia College—our dean—was a widely respected liberal political scholar who was expected to succeed Kirk when he retired.

My friend from the SDS Action Faction, Bob Feldman, a slender, working-class Jewish kid from Brooklyn who had originally uncovered the IDA connection, was already burrowing into Kirk's filing cabinets, finding juicy evidence of the university's collusions with newspapers and its schemes to keep its research secret. (That week the documents would be published in the underground newspaper the *New York Rat*.) Other students were trying out his chair, smoking his

White Owl "President" cigars, tasting his sherry, and checking out the contents of his drawers.

I wandered over to Kirk's magnificent mahogany desk, sank down exhausted in his executive armchair, and idly picked up the phone. The dial tone was still working and, without thinking, I called home. My father answered, still drowsy from sleep.

"We took a building," I said.

"Well, give it back," he replied from our ranch house in Maplewood, fourteen miles outside New York City.

My mother was on the extension. "We've been worried sick," she chimed in. "We saw on television about the takeover of Hamilton Hall yesterday. We're worried you're going to get hurt or arrested."

"I'm okay, Ma," I said. "It's just something we have to do. This war has been going on too long."

"Please be careful. I don't want you to get in trouble," she said.

"I think I may already be in trouble. We haven't seen any cops yet, but they're coming. I can't talk more now. See ya later."

I've often wondered, over the years, why I called my parents that morning. I couldn't have been seeking their approval, something I knew I'd never get. Maybe it was simply that Jewish boys call home, it's that deeply ingrained.

Or maybe I was unconsciously bragging to them: "Look, I'm finally an American, just like you wanted. I'm not afraid to stand up, to speak out. I don't need to keep my head down, like you've always done. I can protest like any other American. I've made it in America! Oh, by the way, you may not be crazy about this part, but I'm going to work to bring down the dumb goyim who run this country. I'll be the Jewish defender of the weak and the downtrodden."

I hung up. For the last few minutes, rumors had been floating around that the New York City cops were downstairs in the security office, ready to bust us. What to do?

Per SDS custom, we convened a meeting in the hallway outside

Kirk's office, at the side of the large rotunda. Discouraged and bone-weary, leaning against one of the columns, I began a stream-of-consciousness monologue to the fifty or so people there.

"No one on the outside must ever know what we say here. The reason we were asked to leave Hamilton was that we weren't solid. I didn't want to tell you this before, but the blacks have guns and are prepared to make a stand; I'm not.

"I'm not ready to sacrifice my life. There are still things I want to accomplish, and I didn't want any of my people to get hurt. That's their fight; we have our own. For some of us, our academic careers are already ruined. The only thing we can do is to make our stand and try to win.

"I didn't want to leave Dean Coleman there"—as a hostage—"with guns and all that in the building, but I had no choice."

Everyone was quiet, eyes welling with tears. We slowly began to discuss the alternatives. Someone suggested we should barricade the whole building; someone else said we should leave. When it was proposed that we just stay in the rotunda, I pointed out that all the cops had to do was close the tall iron gates that surrounded the space and we'd be prisoners.

"Maybe we should barricade ourselves in Kirk's office," I said. "Or maybe we should all leave and hold a vigil in front of Hamilton to protect the black students," I contradicted myself. "I don't see the point of getting busted here. I don't know what we should do." I sank down, my head in my hands.

It was at this point that Sue, my girlfriend, whom I hadn't seen for quite some time, came up and said, "Mark, you're not making any sense. You're incoherent."

I looked at her and nodded, "I haven't had any sleep for two nights straight. The night before the Sundial rally, I was up all night writing a speech, which I never gave because the whole plan broke down. And last night, in Hamilton, you were up, too."

"How long are we going to stay here?" she asked.

"I don't know."

"Well, it doesn't matter anymore. I just threw my tickets for Nureyev and Fonteyn's ballet tonight at Lincoln Center to Peter and Elaine outside," she said, laughing.

Just then someone shouted that the cops were coming up, and Sue and I raced back into the president's office. Dozens of students were exiting through the windows onto a ledge that led to iron window grates you could climb down to the ground. People were debating what to do. In one room Tony Papert, an SDS member who was the leader of the Maoist Progressive Labor faction, was sitting in a circle with about thirty others arguing about whether and why they should stay.

I ran into the room and broke into the discussion. "We can help the blacks by forming a blockade in front of Hamilton. Why take the arrests here?"

Tony, a smart and determined guy, said, "We ran from the gym site, we ran from Hamilton, let's not run from here."

I wasn't thinking too clearly. At some point I mentioned needing to do "dorm organizing," which made no sense at all.

So Sue and I jumped out one of the president's windows with the others, making what we thought was a timely exit. When we got down to the ground and moved across the lawn out onto the sidewalk, we looked back and saw, through the second-story windows of the president's office, the figures of blue-coated cops in the rooms where we had just been. About twenty-five people, led by Tony, remained behind in an interior room.

The hundred of us who had just jumped out the windows stood around in the rain outside Low for about an hour. But nothing happened; no one came in or out. No bust. We could see cops taking the Rembrandt and the big TV out of the office, but that was it.

Sue eventually said, "I think we should get some sleep," and we walked out to West 115th Street and her apartment. I closed my eyes the instant I lay down. My thoughts were a blur: *What was the situation we were in? What was going to happen? What should we do? Was I in over my head?* Within seconds I was asleep.

Four hours later I walked back to campus, groggy, for an SDS meeting to be held at Ferris Booth Hall, the student center. First I went to Low to see what was happening. Evading the police guards outside, I climbed through a window and found people in very high spirits, settling in for a long occupation. The original twenty-seven had been joined by dozens more, and they were organizing themselves into committees, including sanitation (there was exactly one bathroom, which people lined up to use) and food. After chastising me for jumping out the window earlier that morning, the Low occupiers ordered me to go to the SDS meeting and recruit reinforcements. I agreed.

About seventy-five SDSers were meeting at Ferris Booth Hall to decide what to do. The old leadership, consisting of the two Teddies, Kaptchuk and Gold, joined by SDS veterans including David Gilbert, now a graduate student at the New School downtown, argued for education and dorm meetings that night in order to increase the base of support among students.

I was disgusted. When it was my turn to speak, I proposed joining the occupation of Low. "The black students and community are in Hamilton, several dozen whites have taken the president's office, and you talk about holding dorm discussions. The time to act is now: Let's reinforce Low!" A vote was taken: 70–3 against my proposal.

"All right, let's take another building," I proposed. This was voted down by the same margin. Enraged, I screamed, "I resign as chairman of this fucking organization!" and stormed out of the room.

As it happened, neither the SDS debate nor my resignation made any difference to events (nor did anyone in SDS take my resignation seriously). The situation was way beyond control by "leadership." Students were joining the Low occupation on their own. Architecture and urban-planning graduate students—particularly sensitive to the needs of the community surrounding Columbia—

seized Avery Hall, the School of Architecture building, in support of the six demands. This was a big boost, because SDS didn't even know these people.

All Wednesday, visitors and supporters streamed into Hamilton Hall to offer support for the black students there. Donated medical supplies, food, and volunteers came from Harlem CORE. Along with other civil-rights and community groups from Harlem and the rest of the city, Rap Brown and Stokely Carmichael of SNCC, which had been championing black power, met with the students in Hamilton and then held an angry press conference in front of the building, denouncing Columbia's racism. "If they build the gym, burn it down," Rap Brown said. Large demonstrations of black high-school kids and other community people spontaneously formed on Amsterdam Avenue, next to Hamilton.

By 3:30 P.M. Wednesday, the Hamilton students asked Dean Coleman and Proctor Kahn if they wouldn't like to go out to get some dinner. They probably had become aware of the danger of potential kidnapping charges and also realized that holding hostages was detrimental to the cause. It had, after all, been a spontaneous act, taken in the heat of the confrontation the day before, and had not been completely thought out. Both men went directly to an emergency faculty meeting then under way, and their dramatic entrance received a standing ovation.

Early the next morning, humanities and social-science graduate students numbering about two hundred people established the Fayerweather Commune in what had been Fayerweather Hall, a classroom building. Four buildings were now occupied, all unplanned.

On Wednesday night a university-wide strike was declared at a mass meeting in the student center; the next morning pickets went up at most classroom buildings. That morning, while walking a picket line in the rain, I found myself face-to-face with Vice President David Truman, who was demanding access to the university's computer center. "I'm sorry," I said, "the university is shut down by order of the students and community." He shot me a disgusted look, murder flashing in his eyes, then stalked away with his entourage.

By Friday the administration was forced to close the university, because the situation was completely beyond their control. In the early hours of that morning a guerrilla commando force of thirty, armed with a hacksaw and led by JJ, had seized a fifth building, Mathematics Hall, in order to ease the pressure on Low, which was filled beyond capacity. Math Hall promptly filled up with two hundred people, both students and nonstudents, including the Motherfuckers from the Lower East Side, Abbie Hoffman of the Yippies, and Tom Hayden, a past national president of SDS and author of the 1962 Port Huron Statement. Tom wound up chairing the Math Commune meetings in a terrific display of SDS's participatory democracy: "We've heard from the side urging going passive when the police come. Now, is there anyone who wants to speak for a more militant position?"

The longer we held the buildings, the more people joined us.

A strike coordinating committee (SCC) was established, with two representatives from each building, plus two from Strike Central, located in the student center. I was elected chairman. The blacks in Hamilton Hall were invited to participate, but they declined so they could act autonomously as representatives of the Harlem community. It made sense that they would want to be masters of their own fate. Periodically during the six-day occupation, however, they assured the SCC they would not accept amnesty only for themselves but would hold out for the six demands.

Strike Central was my base of operations. I collaborated constantly with SDSers David Gilbert, Ted Gold, and Ted Kaptchuk, as well as two people I had never worked with before, Lewis Cole and Juan Gonzalez, plus many others. Juan, clear-spoken and judicious even though fired up in a nonthreatening way, had come from the Citizenship Council, the liberal "student activity" that had organized tutoring in Harlem, among other programs. Having grown up in Spanish Harlem, he was one of the few Puerto Rican students at Columbia. He liked to say that if all the Latinos at Columbia wanted to hold a meeting, they could have done so in a phone booth. Lewis, habitually hunched over because of his height and lankiness, with a shock of curly black hair, was a theoretician and intellectual who loved to hear himself speak, later sometimes described by the press as "the brains behind Mark Rudd." I adored them both.

The energy was electric. We held SCC meetings, arranged for physical support—about a thousand people had to be fed—collected money, talked with reporters, put out flyers and pamphlets explaining the strike. At one point Strike Central had three rented mimeo machines operating simultaneously, state-of-the-art Gestetners, which in those pre-Xerox, pre-digital days cranked out pages by the thousands, around the clock. We had a bank of new phones connected to take calls from supporters and press around the world.

People at Strike Central typically worked eighteen- and twenty-hour days, as you would in combat. *Sleep? What's that? Here, have a cup of coffee and a Danish. A group of peace ladies from Queens just brought in a huge donation of home-cooked food.*

We grew close the way any fighting unit does—soldiers, firefighters, cops, construction workers—sharing the necessary work. Strike Central didn't consider ourselves a collective in the same way as the communards in the occupied buildings did, alas. There was a difference.

The atmosphere in the occupied buildings was charged even higher than in Ferris Booth. A strange, chaotic, and loving new life-form had spontaneously erupted. Students normally isolated in their apartments and dorm rooms, or held hostage by curfews in the women's dorms, were now living together in groups numbering up to 250. The term "commune" was joyously seized upon, both as a historical reference to the short-lived Paris Commune of 1871 and even more so as a symbol of a new, collective future.

The buildings were ours—at least until the cops came. We had to figure out food, sanitation, security, all the conditions of life. Absolute strangers started talking to each other, intensely and intimately, about what was happening, about our fears of the coming police bust and for our futures. This latter concern led to discussions about the meaning of our action for our lives. Living on the edge like this required constant talk, talk, talk.

Then there were the endless meetings about the six demands, about negotiating with the administration and the faculty, about what to do when the cops came. Fayerweather, for example, debated the question of amnesty for several days nonstop, sending representatives to the SCC periodically with their latest, sometimes contradictory votes. Was it more consistent with "civil disobedience" to take punishment? Did amnesty detract from the other demands? Should we let it go? Or did amnesty symbolize the fact that we were right and the university was wrong?

In the more radical buildings—Low and Math particularly—amnesty symbolized the rightness of the cause. I argued for it perhaps hundreds of times in the occupied buildings, at the Strike Steering Committee, to the faculty, in a few informal negotiations with the administration, and to the press.

I happened to be in Low Commune (the building formerly known as Low Library) during a debate on how to respond to the police attack that everyone knew was eventually coming. Low was a "hardcore" building, meaning that the occupiers never wavered on the question of amnesty, as Fayerweather and Avery sometimes did. There was difference of opinion, however, on whether the occupiers should barricade and fight the cops, resist passively by going limp, or accept arrest and walk out under their own power.

At one point JJ proposed that they place President Kirk's Ming vases on the window ledges to deter the police. People were aghast at the possibility of physical destruction—we were no Red Guard from the Chinese Cultural Revolution, after all. JJ then modified his suggestion. "Okay, let's just push the cops off the window ledges nonviolently," he said as everyone laughed.

In the end each group decided that its members would gather in a separate room to confront the police as they wished. When the bust finally came, days later, the cops made no distinction between those who resisted, went limp, or cooperated; all were equally and arbitrarily beaten with clubs, punched, and kicked by New York's Finest.

Despite the anxiety over the ever-present danger of a police bust, I heard people express, over and over, the same feeling that "finally I'm doing something with my life that counts, that can make a difference." Liberated people in liberated buildings had broken out of the mold of passive student. This was the critical element of the communes that we didn't have in Strike Central: the feeling of transformation and empowerment that comes with being part of a group democratically deciding its own fate. Banners went up on the buildings: LIBERATED ZONE, JOIN US, VIVA LA HUELGA! and WE SHALL NOT BE MOVED!

Along with endless meetings, the liberated buildings became host to continual celebrations of freedom. One evening a performance in Fayerweather by the guerrilla theater troupe the Pageant Players led to raucous drumming and dancing, which in turn led to the spontaneous wedding of grad students Andrea Boroff and Richard Eagan, calling themselves Mr. and Mrs. Fayerweather. The Reverend Bill Starr officiated, pronouncing them "children of the new age."

Outside the liberated buildings, the campus was in total turmoil. A "Majority Coalition," deliberately misnamed, had formed around a nucleus of jocks and conservative students. They attempted to stop food and reinforcements from reaching the occupiers of Low. At one point, serious-faced faculty "monitors" placed themselves between the opposing lines; they were supposedly there to keep order, though it looked to us as if they were actually reinforcing the jock lines. What was the difference between a jock intercepting a loaf of bread in the air and a professor doing the same?

Fistfights and shoving matches occasionally broke out, though more often the radicals and the jocks just hurled verbal taunts and jeers back and forth. They called us "pukes." In turn we delighted in pointing out to the jocks that no Columbia line had ever held, a clear reference to the dismal record of the football team even then (Columbia currently holds the college record for most games lost).

We never took the jocks very seriously, despite their threats to throw us out of the buildings. Not only were their numbers small—you saw very few blue armbands around campus—but to us they were partisans of the dying, stodgy past who were defending an indefensible status quo. We were what was happening; we had the future—the revolution—on our side.

Much more numerous than the jocks were the group of a thousand or more people known as the "Green Armbands" who were against violence and supported our demand for amnesty. Those of us occupying the buildings, the radicals who wore the revolutionary red armbands, looked on them with contempt as liberal fence-sitters. In actuality they were our closest allies and were merely unwilling themselves to join the occupation. When the bust came, they bravely placed themselves in front of the entrances to the occupied buildings to prevent the cops from getting in. Many of them were severely beaten and injured.

The uproar equally engulfed the Columbia faculty. The reaction of senior faculty, the stars who were the most privileged and conservative in the school, tended toward indignation: "How dare you occupy our buildings and interrupt our brilliant teaching?" they sniffed. The renowned professor Lionel Trilling, who had once been a leftist, warned that SDS should not be allowed to turn Columbia into "some scruffy Latin American university."

But a few of the liberals among the senior faculty and a larger number of associate and assistant professors—known as "junior faculty"—tried to mediate between the students and the administration. Many of them were young instructors who were sympathetic to the political goals of the occupation, if not our tactics. They also had their own grievances against the administration, such as lack of tenure, low pay, lousy housing. Neither Kirk and Truman nor the trustees would budge on the Ad Hoc Faculty Group's compromise proposals. Nor would the Strike Committee; we believed that by holding fast to our demands we could only gain support.

By talking with the faculty, we did gain time. On several occasions they were successful in convincing the administration to postpone a bust because negotiations were in progress. They opposed the administration in its refusal to yield on the gym issue, which many faculty were more than happy to concede in order to defuse the crisis. The Ad Hoc group was also rankled by the administration's refusal to yield any of their disciplinary power, which they saw as both an obstacle to negotiations and a negation of the faculty's authority.

No more than a handful of professors and instructors—men such as Richard Greeman, Terry Hopkins, and Immanuel Wallerstein—understood and appreciated the issues we were raising in the six demands or the necessity to occupy buildings to win those demands. They were leftists who understood our visceral revulsion to the militarist and racist university. Unfortunately, the vast majority of faculty had no empathy with us or our politics and believed that we were "destroying the university," as if it were a fragile plant that wouldn't survive this crisis.

At an early Ad Hoc Faculty Committee meeting, sociology professor Alan Silver asked me, "Doesn't the university have any redeeming features that merit your saving it before that chain [of events which SDS began] leads to disaster?" At that moment I didn't answer, because I didn't want to be seen as putting the need to build an antiwar movement—through confrontation around larger issues—above the needs of the university. But the truth was that ending the war did mean attacking the military-industrial university: SDS had always wanted "a free university in a free society," as one of our slogans read. I should have answered Professor Silver with a question of my own: "Can't you see that there's a point where the status quo becomes intolerable?"

The administration also held tight to its position of no negotiations, with the exception of trying to talk with the black students in Hamilton Hall, who were offered separate amnesty. At the beginning of the occupation, Kirk and Truman had tried to call in the New York City cops to arrest the white kids in Low Library only, while leaving the Hamilton Hall blacks undisturbed. The police official in charge said that he couldn't bust one group of trespassers and not the other.

The Hamilton Hall students did not accept the administration's offer, sending word to the Strike Coordinating Committee that there would be no separate deal on the six demands. The de facto coalition held.

Along with all these campus factions—the occupiers, the Green Armbands, the jocks, the administration, the junior and senior faculties—people from Harlem and other parts of New York City streamed onto campus. Hundreds of students from other campuses, including high schools, came for support rallies or joined the occupation. Scores of young black kids, attracted by the action, were running around the campus. Not allowed into Hamilton Hall, they were accepted in Fayerweather, the most open commune. Day after day local black leaders led demonstrations of hundreds of community people in support.

These loud and aggressive demonstrations particularly worried Kirk and Truman. As President Kirk wrote in his "Message to Alumni, Parents, and Other Friends of Columbia" later that summer:

> There was a natural and serious concern at city hall, which we shared, about the possible reactions of our neighbors in Harlem to the use of police in Hamilton Hall. As indicated, Hamilton was entirely occupied by black students, and, in view of what happened in Harlem following the assassination of Dr. Martin Luther King, the possibility of actions directed against the university had to be weighed carefully.

The support from the Harlem community was as great a source of morale for us in the occupied buildings as it was a threat to the administration. All five buildings were continually supplied with food that community members donated or cooked. The Hamilton Hall students many years later told us that they had never eaten so well in their whole time at Columbia; the ladies of Harlem had made sure that they had home-cooked meals. From a window of Math Commune, I saw a city bus stop out on Broadway, the black driver get out and put a donation into a bucket lowered down from a ledge. Through hundreds of small gestures, the New York City community made it clear to us that we were involved in something much larger than ourselves and our elite university.

The administration hesitated, the faculty tried to negotiate, the jocks blubbered that nobody was doing anything to save their university from the radicals, and we held tight.

The situation could not last.

5

Police Riot: *Strike!*

The long-awaited bust finally came in the very early morning hours of Tuesday, April 30, almost a week after we first seized Hamilton Hall. I first got word of it the day before, in a meeting at Strike Central with two of Mayor John Lindsay's representatives, Barry Gottehrer and Jay Kriegel. "I'm sorry," one of them said to me and Ted Gold, "but the Columbia administration has decided they are calling in the police, now. We didn't want this to happen, but we can't stop it." I wasn't surprised, but at the same time it was difficult to accept that the end was finally near.

"Thanks for your help," was all I could say as we shook hands and quickly parted. Ted Gold and I needed to pass the word to the Strike Committee and the occupied buildings.

Mayor Lindsay was a liberal Republican who was no fan of Columbia's taking over the city park to build the gym. He sent Gottehrer and Kriegel on the first day of the occupation, April 23, in an attempt to mediate between the administration, the students, and the police. Both were Jewish and not that many years older than I. I could relate to them as older brothers; they spoke directly and projected sympathy. Or maybe they were just really good at their jobs and I was taken in. It didn't matter either way, because neither the administration nor the strikers would budge.

After nightfall we began to hear reports at Strike Central of police vans and buses moving slowly through the surrounding streets. Dimly lit battalions of hundreds of uniformed, helmeted police were massing in the night, armed with billy clubs and truncheons, as the machine prepared to strike.

Inside the five liberated buildings, people prepared for the attack by reinforcing barricades of chairs and tables and distributing wet rags to be used against tear gas. No one knew what would happen; the fear and tension were later described to me as overwhelming. Finally the endless meetings on what to do when the police came were over, as a thousand people in the buildings waited, some sitting quietly on the floor, others singing civil-rights freedom songs. Outside the buildings hundreds of people wearing green armbands formed themselves into ranks in front of the entrances, hoping to block the police.

As the iron gates to the campus closed, crowds began milling outside on Broadway, howling, "NO VIOLENCE, NO VIOLENCE!" That's where I was, along with Lew Cole, Juan Gonzalez, and several others from the SCC. The Coordinating Committee had decided that we should not be arrested so that we would be available to organize the mass strike we knew would result the next day. Besides, we weren't commune members, occupiers of the buildings; we had served in support roles on the Strike Coordinating Committee.

Above the roar of the crowds, we could hear people screaming and crying. On campus, phalanxes of police had suddenly burst through the cordons of pacifist Green Armbands, nightsticks and blackjacks flailing, wielding crowbars to break in to the barricaded buildings.

Each commune had chosen its own tactics, ranging from barricades and passive resistance to cooperation with the police by submitting to arrest. Sometimes, as in Low, when no unified consensus could be reached, different tactics coexisted in the same building. Everyone who was willing to submit to arrest gathered in one room, while those who wanted to resist nonviolently by going limp were in another. In the end the police randomly, arbitrarily beat those who cooperated as well as those who resisted. In some cases all the demonstrators from a certain room were made to run a gauntlet of cops who beat them with clubs and blackjacks. Elsewhere people were led quietly and professionally to waiting vans.

Some of the worst police violence was directed not against the occupiers but toward the people assembled outside the occupied buildings. Police attacked the Green Armbands, faculty members, bystanders—even the Majority Coalition jocks who were there to cheer them on. Surprise! Clearly identified medical volunteers were beaten as they tried to help the injured. At one point the police chased several hundred people toward a locked exit between two dormitories, beating with nightsticks and blackjacks everyone they caught. Students attempting to seek shelter in dorms were chased inside and beaten in the lobbies and stairwells. The police went berserk rioting that morning. Blood was everywhere, on the steps of the occupied buildings, on the herringbone brick walkways I had loved so much as a freshman, even in dormitory lounges.

The next day I read that New York City police commissioner Howard R. Leary commended his force for handling a "potentially difficult situation without a single case of serious injury." What? History professor James Shenton was knocked to the ground, kicked and beaten repeatedly, as were Rabbi A. Bruce Goldman, Jewish chaplain of the university, a *New York Times* reporter, and a number of other members of the faculty. The pictures of dazed students, blood streaming down their faces, tell the story much more accurately than the commissioner did. St. Luke's Hospital, across Amsterdam Avenue from Columbia, treated eighty-seven students, while midtown Knickerbocker Hospital got the more seriously wounded. In all, several hundred people were injured, including about thirteen cops, and 720 people were arrested.

What fueled the police mayhem was probably a combination of the cops' right-wing political anger about our opposition to the war and their class resentment toward what they perceived as privileged middle- and upper-class Columbia students. The cops' children didn't go to Columbia. Perhaps there was some anti-Semitism, too, on the part of the predominantly Irish cops toward the radical Jewish kids. It certainly didn't help that we antagonized the cops by calling them "pigs" and "motherfuckers." Unfortunately, we believed that as agents of the enemy (the ruling class), they had become the enemy. Most of us weren't too strong on our nonviolent philosophy. It's a good thing the Black Panther slogan "Off the Pig!" hadn't reached New York yet, or some of us might have been murdered that night.

Beside myself with anger and despair, I was pacing in front of the Chockfull o' Nuts coffee shop across the street from the Columbia gates. The sirens were wailing, and thousands of people were screaming, as if in unison. One of my professors, David Sidorsky, a conservative Jewish intellectual who taught me the first rule of positivist philosophy, that "A is A," appeared out of nowhere and screamed into my face, "Look at what you've done, look at what you've done!"

Equally out of control, I screamed back at him, "You've done it to yourself! You could have supported us, but all you care about is teaching your classes!"

"Isn't that what I'm here for?" my freshman Contemporary Civilization professor asked.

"And while you ignore the war, the university helps carry out mass murder in Vietnam."

"I won't stand by while you and your friends destroy the university!" he screamed at me.

"It looks like the administration and the police are destroying it right now." We could hear the wails from inside the gates.

Down Broadway somebody threw a trash can through the window of the Chemical Bank branch on the corner of West 113th Street. The connections between business, finance, the war, and the violence of the state seemed to come together, at least in my mind at that moment, as the glass crashed to the pavement. Breaking away from Professor Sidorsky, I ran down the street, picked up a brick I saw lying around, and, in a puny gesture, shattered the post-office window next door. Throwing that brick gave me no solace.

While all this mayhem was raging on and off campus, a carefully supervised contingent of police entered Hamilton Hall through an underground tunnel. They first dismantled a barricade of tables and chairs the black students had built, then went upstairs and arrested the eighty-one occupying black students. The SAS leaders had previously asked all community people to leave and not take the bust. The students were led out again through the tunnels to waiting police buses and taken off for arraignment. No one was injured. We at Strike Central later learned that the bust had been negotiated between the Hamilton Steering Committee and the police.

No one from the police or the city or the Columbia administration ever approached me or any other members of the Strike Coordinating Committee to work out a deal for the peaceful surrender of any of the other four occupied buildings. I suspect they reasoned that we were no threat because we didn't have Harlem behind us. Perhaps we were thrown to the tender mercy of the cops as a reward for their having waited a week cooped up in buses, gnawing on their nightsticks. I never heard anyone complain about our unequal treatment, however, nor did anyone resent SAS for having negotiated a surrender. The position of the black students was perfectly understandable.

The next morning the arrested students began straggling back from the downtown detention centers and courthouse, tired but undefeated. Their anger was mixed with pride over what they had endured. Everyone was still amazed that their university administration had attacked its students with such violence.

That afternoon from a window ledge of Hamilton Hall, perched high above Amsterdam Avenue, I looked over a crowd of more than a thousand students, many of them fresh out of jail. They were joined by hundreds of community supporters. It was just twelve hours since the cops had invaded Hamilton to arrest and punish us.

"We won't let them turn us around!" I screamed to the crowd, and the crowd roared back, "STRIKE, STRIKE, STRIKE!" as they held up their arms in the V sign that had become the symbol of the strike, the same sign used by the antiwar movement to demand peace in Vietnam. I suddenly recalled a dream from years before, about my standing on the top of a building addressing a crowd just like this.

I noticed then that a contingent of thirty-five uniformed cops began pushing through the crowd to reinforce the twenty-five already at the Amsterdam Avenue gate. Several hundred students who had been rallying inside the campus, at the Sundial, rushed toward the gate, yelling "COPS MUST GO!" and linked arms to block them. The bluecoats then charged the line, and a melee broke out, with students punching cops, cops jumping students and beating them to the ground with clubs, then kicking them for good measure. A student jumped feetfirst from a second-story window ledge onto a cop, seriously injuring him, at which point the student was surrounded and beaten by plainclothesmen. Blood splattered the walls of classroom buildings. The crowd outside the gates surged forward to see what was happening.

Desperate to stop the bloodshed, I climbed onto a window grating and told people that "the way to win is not to go out and fight the cops." The deputy inspector in charge, his white gloves stained with blood, then ordered his men to pull back. According to the editors of the *Spectator,* "A police official told reporters later that no policemen had received orders from Columbia officials to clear College Walk and that they had not been authorized to use clubs." (Forty years later I found out that the injured cop, Frank Gucciardi, had suffered a spinal injury that debilitated him for the rest of his life.)

That night, Wednesday, May 1, students, faculty, and community supporters packed Wollman Auditorium, the largest auditorium on campus. Somehow 1,300 found room in a space with a normal capacity of 750. The Strike Coordinating Committee, which had previously represented only the people occupying the five buildings, had called this mass meeting to reconstitute itself in order to include the new strike supporters who were joining us by the thousands. We had to answer these questions: Would we keep to the original six demands? Would we try to transform the university now that the administration had exposed itself as violent and morally bankrupt?

For two hours speaker after speaker angrily denounced the administration for its use of violence, and then, at 10:00 P.M., David Gilbert rose to the chair and began the night's real business—restructuring the Strike Committee. In his soft-spoken manner, he told the crowd that democratic procedures would be difficult but that all of us in the room could "try to participate on a large scale in making fairly complicated decisions. . . .

"The original six demands are no longer sufficient," he said. "In addition to winning political demands, we must begin to create a new university." The crowd cheered wildly.

Earlier that afternoon the old Strike Coordinating Committee had met and decided that Tony Papert, who had a lot of prestige because of his solid leadership in Low Commune, and I would present a proposal to expand the committee by first ratifying the original six demands. Then we could move on to talk about remaking the university.

In my remarks to the Wollman crowd that night, I began by admitting that on the first day of the occupation I hadn't known what to do, "whether to go to McMillin Theatre to talk to Truman, whether to try to get into Low Library, whether to go to the gym site, what to do. But we did them anyway. All along the line, people said, 'Hold back, you're really going to turn people off, because your radical politics and especially your radical tactics are just no good.' The fact that there are about a thousand or more people in this room . . . is proof that the sort of politics and tactics we offered were right. . . .This is almost a fact of the strike."

I then described a conversation I'd had during the occupation with a professor who warned me that if there were a bust, everything would be lost. "Well, we got busted, and now look around you," I said, smiling.

People screamed, "STRIKE, STRIKE!" in response.

"Let's not be timid; let's keep pushing," I said. "Let's be extremely clear on what we're demanding, and let's be clear in why we're doing this. . . .We're doing this to create a human society and to fight exploitation of man by man, and we think that this university was an example of this exploitation."

By this I meant, *If you want to join the strike, welcome, but don't try to change what we're about.*

Tony then proposed we use this moment, while the administration was weak, to take over the university's functions. "We should form a provisional government," he said.

The moderates—whom we called "liberals"—in the crowd were fuming. A student representing an ad hoc group of 250 graduate students rose and presented an alternate plan: Only support for the strike would be necessary for participation, not reaffirmation of the six demands, to ensure that the "coordinating committee represents as broad a spectrum of campus opinion as possible."

For the next two hours, both sides went back and forth, passionately arguing the positions. The moderates wanted to remove the cops from campus, obtain the resignations of those in the administration responsible for the decision to use force, and encourage student participation in restructuring the university. The radicals wanted to keep stressing the six demands with their focus on the university's complicity in the war and its racism and also to push for a more thoroughgoing seizure of power.

Finally David Gilbert, calmly chairing the meeting, called for a voice vote over the two proposals. A clear majority of those present voted for the radical Strike Committee plan. But Tony and I didn't want a split with the moderates—it would weaken the strike. We conferred briefly, and Tony told me not to bother with the vote and to accept the moderates' proposal. He later regretted the advice. Without a pause for reflection, possibly because of my weariness, I grabbed the microphone, turned to the audience, and said, "We accept the graduate students' proposal."

The room erupted in deafening applause and cheers. My unexpected move—I surprised even myself—had averted a split on that first day of the expanded strike. Sympathetic professors rushed up to me, shaking my hand, saying they'd never seen such a brilliant political maneuver. But even at that moment, I suspected that the inevitable split between the liberals and the radicals was only postponed.

The next day a reconstituted Strike Coordinating Committee met and reaffirmed the six demands, which meant that my spontaneous gamble had worked. Two co-chairs were elected, me and a "liberal." Within days the expanded Strike Committee grew to seventy delegates, each speaking for seventy signed constituents, for a total direct representation of about five thousand people. The militant core of the support was still the radical occupiers of the buildings, but thousands more flocked to join after the bust—graduate students, community people, faculty, university employees, all were welcome. Many grief-stricken professors who had previously tried to negotiate with an intransigent administration publicly joined the strike. Even a fraternity sent a delegate.

The strike lasted another month, until the end of the school year.

The university administration attempted to reopen classes after a week, but our picket lines around key classroom buildings were respected, and very few regular classes met. "Liberation classes" were held on the lawns or in professors' apartments, as authorized by an entity of the SCC called the Strike Education Committee. You could now take courses like History of the Spanish Student Movement, Political Aspects of William Blake, and Columbia and the Warfare State.

Walking across campus in those weeks was like navigating through a revolutionary jamboree, with classes meeting under trees, guerrilla-theater groups performing, music constantly being made, including a free performance by a small young band out of Palo Alto, California, called the Grateful Dead. People greeted me as if I were the mayor. I remember even kissing a baby whose proud parents thrust him into my arms; that's still embarrassing.

Political caucuses met continually—graduate students, special-interest groups, SDS, the communards, liberals for university restructuring, junior faculty, senior faculty. Everyone talked constantly about the nature of education and the society, about revolution, about defining our goals and priorities. The weeks right after the bust were a florescence of energy and imagination such as Columbia had never seen.

A group of senior faculty even tried, timidly, to seize power from the administration. The Executive Committee of the Faculty met directly with the Columbia trustees, a first in the two-hundred-year history of the university, and found that the people who decided policy knew nothing at all about the causes of the strike and the students' demands, having been informed by Kirk and Truman that only a small bunch of troublemakers were responsible. However, the corporate guardians were not about to turn the keys to their university over to a bunch of lowly academics.

On the Saturday afternoon after the bust, a well-meaning senior faculty member, professor of psychology Eugene Galanter, who had been active in organizing the faculty to mediate between the administration and the occupying students, called a secret meeting in his beautiful Riverside Drive apartment overlooking the Hudson. Those of us on the strike side arrived together—Juan Gonzalez, Ted Kaptchuk, and myself—plus graduate student Ed Robinson, the liberal co-chairman of the SCC.

Professor Galanter graciously seated us at his polished cherry-wood dining table. He poured tea for each of us from a silver tea set. We exchanged a few pleasantries, which the professor took to be an invitation to lecture us on his views of the strike, something about salivating dogs and rat pellets, his psychology stock-in-trade. We had long since learned to ignore most of what faculty had to tell us about the situation.

Suddenly in walked David Truman, vice president of Columbia University and the individual who likely had made the call to bring in the New York City Police Department. He was a small man, balding and graying, with deep-sunken eyes framed by prominent eyebrows. Even on Saturday afternoon, he wore his uniform of brown suit and tie. He looked neither happy nor well. Before the strike, Dean Truman was the heir apparent to President Kirk; now, day by day, he was watching his grand prize in life slip slowly beneath the waters. He dutifully shook hands with each of us, hesitating slightly when he got to me.

Dean Truman methodically went through our six demands, explaining why each one could not be met, especially amnesty for all those arrested and disciplined. We had heard these arguments many times before. At one point, without thinking much about it, I took off my heavy work boots and socks; my feet hurt, and they were also sweating. Perhaps being forced to sit down with the barbarian in the palace made Dean Truman uncomfortable, I don't know.

Ed Robinson, true to his liberal politics, attempted to reason with Dean Truman. I cut him off, saying, "Look, we're leftists and we want to fight the policies of this university."

Truman gave me a disgusted look, as if I were confirming what he already knew, that there was nothing to talk about. This inspired me to spice things up.

"The Strike Coordinating Committee has set up a provisional administration, and we're prepared to run the university. All you need to do is give us the bursar's office so we can pay people."

This suggestion provoked from Truman perhaps the sourest look I've ever received, one of pure hatred. The meeting was over; we didn't shake hands as we got up to leave.

A few weeks later, Vice President Truman told the *New York Times* reporter how he felt about me, personally:

> He is totally unscrupulous and morally very dangerous. He is an extremely capable, ruthless, cold-blooded guy. He's a combination of a revolutionary and an adolescent having a temper tantrum. No one has ever made him or his friends look over the abyss. It makes me uncomfortable to sit in the same room with him.

And all along I thought I was such a lovable boy.

The press attention to the Columbia strike was unforeseen by me or anyone else in SDS, but it was not unwelcome. Our goals went way beyond issues involving only Columbia University, and if we could use the press to spread our ideas and present an example to students at other colleges, so much the better. With almost 8 million young people attending postsecondary schools in 1968, the potential for antiwar and antiracism protest was enormous. The strike at Columbia became linked in the public's minds, quite accurately, with demonstrations and revolts and protests at other schools, within the military, and even with the French and Mexican students out in the streets.

Unfortunately, the actual content of the press coverage was by and large terrible. News stories focused solely on our actions and almost never mentioned our reasons for striking.

In early May, right after the first bust, a widely known TV producer and talk-show host, David Susskind, interviewed me, Lewis Cole, Ray Brown Jr., and Bill Sales of SAS, as well as a couple of guys from the Majority Coalition, on his show. After the panel had spoken, the audience was allowed to participate. The Jewish chaplain of the university, Rabbi A. Bruce Goldman, stood up and told his story of being severely beaten the night of the first bust while trying to help injured students. His arms and head were still in bandages; he looked like a battle survivor in a war movie. Susskind interrupted Rabbi Goldman before he was finished, saying, "Now, Rabbi, it grieves me to see a clergyman of my own faith prevaricating." We couldn't believe our ears. The audience erupted, screaming and shouting against Susskind.

That incident was typical. We tried to explain why we were forced to take action, but the press could only portray us as lunatic, destructive kids. Just one frame of reference was acceptable: university defense contracts and "slum clearance" good, student disruption bad.

The *New York Times* was intimately entangled with Columbia at all levels, from the board of trustees on down, and became the house organ for the Columbia administration. *Times* publisher Arthur "Punch" Sulzberger, a Columbia graduate, was a trustee of the university. Many of the editors were Columbia Journalism School graduates or teachers. My favorite tale of *New York Times* complicity with Columbia was its lead piece in the first edition of the April 30 paper. The newspaper reported that the police had peacefully removed all the demonstrators from the buildings. However, that early edition hit the newsstands *before* the bust had occurred. The *Times* had been tipped off to the plan the afternoon before the bust by police leadership so that the story could be written ahead of time. In later editions the *Times* barely mentioned that one of its own reporters, Robert Thomas Jr., was severely beaten by New York City cops using handcuffs as brass knuckles; that fact didn't square with the story of violent students and peaceable police.

The next day, May 1, *Times* managing editor A. M. Rosenthal wrote a front-page opinion piece masquerading as a news article about the damage the nasty and brutish occupiers of Low had done to President Kirk's office. "How could human beings do a thing like this?" he quoted a cop who was holding up an allegedly damaged book. The investigating Cox Commission, named for former U.S. solicitor general Archibald Cox, later concluded that there was "no substantial vandalism in Low Library." Desks and chairs had not been smashed, and there were no "wrecked rooms" as Rosenthal had reported, just the normal wear and tear of 120 people living for a week in three small rooms.

The *Columbia Daily Spectator* obtained sworn affidavits from professors and others who saw police breaking chairs, throwing ink on walls, and doing other damage to some of the buildings *after* the students were arrested and removed.

Two days after the first bust, on May 2, eighty Columbia students picketed outside Sulzberger's Fifth Avenue home, charging him with a personal conflict of interest and his reporters with generally failing to report what our goals were. Pickets also went up at the home of Willam Paley, head of CBS and another Columbia trustee, whose *Face the Nation* had given President Kirk an entire half hour to present his point of view. We were not invited to rebut or to explain our position on the show.

Editorials in the *Times* and other newspapers, and on radio and TV stations, railed against "hoodlumism" and "irresponsibility over reason," but nobody mentioned the mass murders of Vietnamese by weapons developed at Columbia or the thousands evicted from their homes in Morningside Heights.

When you're involved in events that are so shockingly misreported in the national mass media as the Columbia strike was, you wonder how distorted the reporting might be on other events. From that moment in April 1968, I became suspicious about everything I read and hear in establishment papers or broadcast media.

By contrast, the *Spectator* editors and reporters, all students, led by my friend editor-in-chief Robert Friedman, produced remarkably fair and complete accounts of the strike. They sought to understand the radicals on our own terms, plus looked hard at all the other participants—student, faculty, and administration. Ultimately their reportage was converted into a book, *Up Against the Ivy Wall,* by Jerry Avorn and six other editors of the *Spectator.* It stands out, forty years later, as the best account by far of the Columbia rebellion. WKCR, the student-run FM radio station, also did an admirable reporting job.

The only other media that got the story right was the underground press, especially the *New York Rat,* founded and edited by Jeff Shero, an SDSer out of Austin, Texas, and the Morningside Heights–based Liberation News Service (LNS). The underground papers and LNS were part of the movement just as much as we in SDS were. Their rationale for openly adopting a radical point of view was that the ruling class has its mass media—the TV networks, mass-circulation newspapers and magazines—so the movement needed its own.

Reporters from around the world came to interview me as early as the first day of the occupation. I tried to deflect their questions about my personal life, saying, "The issues are important, not me." But the reality was that I became in the eyes of the press an easy personification of an entire generation of students in revolt: the middle-class kid who had all the advantages, attended an elite university, but who angrily turned against "the system." We tried to put forth other people as spokesmen, but the media sought me out—if they couldn't have Mark Rudd, they wouldn't talk to anyone else. "Mark Rudd" was the archetype of the young, fiery, rude anarchic campus revolutionary who made campus revolt fashionable. In the standard media paradigm, there's always a single male leader, and everyone else constitutes the followers.

Lost was the enormous fact that the organizing at Columbia was the work of hundreds of people at least as committed, intelligent, and articulate as I was. The media didn't know how to represent all the people who had over the previous seven years researched IDA, canvassed in the dorms, attended strategy meetings, written position papers, worked with tenants in the community, occupied the buildings, and endured beatings from the cops. Instead they fixed on one person.

A white person, I should add. With the press coverage focusing on Mark Rudd and SDS, the central role of the black students from the Student Afro-American Society in Hamilton Hall and of community support got submerged to the point of extinction and invisibility. This racist whitening of the strike was probably the greatest distortion introduced by the press, but we whites in SDS didn't even understand what had happened until forty years later when Hamilton Hall veterans pointed it out to us.

At first my friends in SDS and I thought the whole Mark Rudd thing was funny. We batted around ideas for making money for SDS—like my mother and I offering to pose nude for *Playboy* centerfold pictures—but I had a gnawing sense that I was in over my head. I was neither the "revolutionary leader" I was portraying nor the celebrity the media had made me.

My parents were interviewed, too, of course. They cooperated because they thought they could ameliorate their son's bad-boy image. During the occupation a reporter from the *Newark Evening News,* their local paper, called them and got some quotes from my father saying they supported my stand. Identified in the paper as a retired army lieutenant colonel, my father criticized Grayson Kirk as being "inept and senile" and described me as showing "concern for people." A May 19, 1968, Sunday *New York Times* front-page profile of me opened with my mother saying, "My revolutionary helped me plant those tulips last November; my rebel." She was the archetypal doting Jewish mother in the official version of the Mark Rudd story.

They were, as a result, deluged with crank letters and postcards threatening me and attacking them. My mother saved several dozen. Here's a small selection:

> PROUD INDEED. He sure made television—he looks like a goon. A dope addict. He's a snake.

> You all stink. Mark should be shot. You should go back where you came from. Send him back to Cuba and go with him. Communists breed evil in their offspring.

> Such a nerve to keep destroying a college that has been in existence for so many years. You should be ashamed to walk the streets—everyone must be fingering you out but people like you are not ashamed and I hope for the coming New Year God will punish him for all the damage he has done the country. *[From a Jewish writer]*

> Your son's arrogance and your fuzzy-liberal parental doting entitle me under the freedom of the U.S. Constitution, to be utterly nauseated, and in favor of throwing the book at him.

> Jew traitors! Why don't you kikes go back to Israel! *[Written over a picture of my father and me, cut out of the* New York Daily News*]*

My parents were strong enough to withstand the onslaught from strangers and friends. Even more than the crazy attacks aimed at them, they worried for me and my future. How conflicted they must have felt on the day toward the end of May when they received two letters from Columbia University—the first informing them that I had been expelled from Columbia University (no surprise, since a dean had told me the first day of the occupation that I was finished at

Columbia) and the second, addressed to me, a form letter from Dean Coleman of Columbia College, our former hostage:

> Dear Mark:
>
> Although the events of the past few weeks make last semester seem a particularly long distance away, I still wish to congratulate you on your fine scholastic record for the fall term. It is particularly important at this time for us all to take cognizance of the primary function of the College. I hope you take appropriate pride in your academic accomplishment.
>
> Very truly yours,
> Henry S. Coleman
> Acting Dean

I had made the dean's list for academic achievement the previous semester.

On Mother's Day, in the midst of the strike, my parents drove to Manhattan to see me. We sat in their car, parked on Amsterdam Avenue, eating veal parmigiana my mother had cooked for her son the revolutionary.

"Now that you've been thrown out of Columbia," she began the refrain for the tenth time that day, "what's going to become of you?"

"I don't know. I guess I'll just devote myself to working against the war and for revolution."

"But what about your degree, your college education?"

"It doesn't matter. There's something new happening. The country will never be the same again. I can do more and learn more out of college than in."

"But you'll need to make a living."

"I'll work in the revolution," I said.

"What's so wrong with this country? Look how it's provided for us. My father was a peddler, and Jake's father was a tailor. This is a wonderful country," she argued.

"And we're murdering people in Vietnam, and there's Harlem right across the park," I replied, gesturing toward the trees a block away.

"Eat up, you'll need the energy," said my mother. "I wonder if I'll ever be able to laugh again, my heart is so broken."

6

Create Two, Three, Many Columbias

From the beginning, the administration had decided that it would never grant the complete erasure of university disciplinary charges and criminal charges against the students. Given this vindictive atmosphere, it was natural that of the six demands, amnesty grew to be our most important, a kind of touchstone of purity. Those who supported amnesty were in the camp of the radical righteous; those who did not were liberals willing to compromise with evil.

Of our six demands, amnesty was the one that we argued about the most in the occupied communes and in the subsequent Strike Committee. At one point I summarized the underlying logic of amnesty in an interview by declaring, "Since Columbia University exists by exploiting the oppressed of the country and the world, the administration is totally illegitimate and does not have the right to discipline anyone."

But the debate raged: Was it just? Was it cowardly? Could an institution survive with any rules if amnesty were granted?

Shouldn't those who choose to commit civil disobedience take their punishment precisely in order to make their points? Or were we demanding amnesty just to save our own skins? The latter point was easy to refute, because had we compromised on total amnesty, we probably would have received only slaps on the wrist.

At the time I was totally convinced of the rightness of our demand. But I've come to realize that the demand for amnesty was both too complicated for easy definition and, worse, distracted from our real goals. SDS's main intention was educating and mobilizing people around the war and racism—that is, "radicalization." We had been quite clear from the start that we were not fighting for "student power," meaning inclusion of students into the university's governance mechanism; we were fighting for much bigger goals. We feared that "restructuring," as student power later came to be known, would detract from the two critical issues—the war and racism—by introducing a third, less important one. Grayson Kirk was absolutely right when he said we were using the university as a staging ground to fight around "less parochial" issues. What could be less parochial than the war and racism?

We fought bitterly with the liberals who joined the strike after the bust about how much to push the issue of restructuring the university. Over time a significant number of them left the SCC and formed Students for a Restructured University. Meanwhile, amnesty grew to overshadow the war and racism as the central issue, doing exactly what we feared restructuring would. Looking back, I wonder what did it matter whether the disgraced Columbia administration admitted we were right and they were wrong?

The Strike Coordinating Committee advertised a major rally at the Sundial for 6:00 P.M. on Friday, May 17. It was ostensibly to revitalize the strike, whose energy had been slowly waning over the previous two and a half weeks. Picket lines had grown sparse or nonexistent. What no one but the SCC leadership knew was that a community group, the Community Action Committee, inspired by the strike, was planning to occupy a partially abandoned tenement at 618 West 114th Street, a building Columbia was attempting to clear of its occupants in the ongoing attempt to expand the university.

Without telling the crowd where we were headed, so as to not alert the police, we led more than one thousand people to the building, where we rallied for the next few hours in support of the tenants' right to their apartments. The tenants charged that 618 was a typical example of Columbia's tenant-removal tactics: no heat, sickening sanitary conditions, broken front-door locks, falling plaster, complete deterioration. They had asked us to help publicize the situation.

I stayed outside with the other students; the occupying tenants were inside. We made speeches, sang songs like "We Shall Not Be Moved," and generally made lots of noise. "NO MORE EVICTIONS!" we chanted. We sang "When the Revolution Comes," to the tune of "When the Saints Come Marching In." At one point I spotted a university official standing at the back of the crowd. "Hey, everyone!" I shouted from the stoop of the building "Look who's here—the man who made all this possible—that son of a bitch in the gray hat standing on the other side of the street—William Bloor, treasurer of the university. Good evening, Mr. Bloor!" The treasurer was responsible for all the real estate and other business dealings of Columbia. As everyone turned to look at him, he slunk away.

By 11:00 P.M. the university administration had again called the cops. It took hours for them to mass. Hundreds of Tactical Police Force (TPF) troops arrived in their vans out on Broadway, armed with blue riot helmets and shields, sticks. Plainclothes cops dressed in sport jackets, narrow ties, loafers, had infiltrated the crowd, just waiting to pin on their badges. Their pseudo–Joe College attire and their crew cuts made it easy to spot them.

Around 3:45 A.M. the cops started pushing the crowd toward Riverside Drive. They were almost tender with us this time, as compared to the April 30 bust, and were clearly under orders to be careful. "Please move down to Riverside Drive," their captain told us. "No one will bother you."

About seventy of us who had decided to stay in front of the entrance to 618 shouted "Don't go!" to the crowd, but those who didn't want to be arrested allowed themselves to be pushed toward Riverside. We sat down on the building's stoop and its sidewalk in order to block the cops from entering.

"I have been authorized by Columbia University to ask all tenants and their guests to leave the building," announced the business manager of the university. "All necessary precautions for your safety have been taken. The New York City Police Department is here to see to that." We all laughed but stayed put.

We were then approached individually and asked to leave. I said no and was carefully handcuffed to another student and pushed into the back of a paddy wagon. They treated me like royalty. When I was booked at the Twenty-sixth Precinct house up on West 126th Street, the desk sergeant looked at me and smiled. "You must be an Oriental, Mr. Rudd, the way you can relax." The cop who had arrested me was congratulated by the others there. We were all then taken downtown to the Tombs.

In total, 117 people were arrested that morning, both inside the building and outside. Fifty-six were Columbia students, the rest community members.

Early the next morning, my father came in from New Jersey and posted my bond—seventy-five dollars on a charge of disorderly conduct. He waited a few hours in the lobby for me to be released from the holding cell.

Newspaper photographers snapped our pictures as we reunited in the old worn marble lobby. We ran outside and down the street. After a short block, they stopped following us, having gotten what they came for.

My father asked if I'd eaten. "No, just some glop early this morning, but I've got to get back to Columbia right away. We need to organize more," I said.

"You've got time to get something to eat," he replied firmly. We went up to Canal Street to Dave's Luncheonette and ordered corned beef on rye sandwiches. Together we had come in to the Lower East Side many times when I was a kid.

Lieutenant Colonel Jacob S. Rudd, U.S. Army Reserves, was a *macher,* literally a doer, a big shot, a guy who got things done. He was used to being in charge in his work and, to a large extent, at home. But I had him befuddled.

"I just don't understand what's happening," he said to me. "I know that Columbia's involved with the war, I understand about the gym and their being slumlords, but why do you have to risk your future? They're bound to throw you out of school."

I thought about how to answer him. "You've always had three jobs," I began. "You worked for the PX"—the Army and Air Force Exchange System—"you were a Reserve Officer, and you started the real-estate business. You always worked hard, *why*? Because you did what needed to be done. That's how you made it from Elizabeth to Maplewood, by working. Well, I'm the same as you, in my own way: I'm doing what needs to be done."

"Yeah, but I knew to keep my head down," he replied. "During the Depression I had friends from college—you remember Benny Schmirnik?—who were Communists. They thought they were changing the world. Benny and I graduated together from Rutgers as electrical engineers. When McCarthy hit, around the time of the Korean War, Benny lost his job with Dumont, the radio and television company, and could never find work. The FBI hounded him. He's broken. I don't want to see that happen to you. It's a dangerous world out there. I never went for politics."

I had heard about Benny Schmirnik often in my childhood. He was an object lesson. "The Communists never built a base," I pontificated. "We're going to have millions on our side. It will take some sacrifices, but somebody's got to be out in front, you know that." I was appealing to the infantry officer in him. The U.S. Army Infantry slogan is "Follow Me!"

Of all the brothers and brothers-in-law in the family on both sides, only Jake had become a financial success. He was a gambler of sorts, a smart one, willing to borrow money from banks and from friends to buy buildings, always calculating in order to make the big deal. The other men in the family had been content to be small shopkeepers, but Jake had had the guts and self-confidence.

"I had to make a living," he said. "You've got the luxury of not worrying.

"I just don't know. I can't say you're making a mistake, because you're caught up in this now and you have to stay with it. It's a bad situation."

We said good-bye out on Canal Street. He got in his car and drove into the Holland Tunnel for home, while I caught the uptown #1 train back to Columbia.

On Tuesday, May 21, just three weeks after the bust, three other SDS leaders and I were officially suspended from school. We had been called into the dean's office to answer charges, but we declined to go, on the grounds that we wanted complete amnesty for everyone involved in the demonstrations.

The strikers felt obliged to take direct action again, this time against the university's attempted repression of our movement, their picking off of our leaders. That afternoon I made a Sundial speech to about five hundred people, saying, "I've got that old déjà vu feeling," after which we marched into Hamilton Hall, not to see the dean but to occupy it. April 23 was being replayed. The university called the police again, and late that night 120 people were peacefully arrested, including seventy Columbia students and a mixed bag of around fifty faculty, parents, community people, and students from other schools.

The police entered the basement of Hamilton through the tunnels, as they had before, and carefully disassembled our barricade. A long-haired plainclothes cop in bell-bottoms and cowboy boots arrested me and put me into a squad car. The next day I read in both the *New*

York Times and the *Daily News* the completely bogus story that this cop, a guy named Ferrara, had infiltrated SDS undercover and had become my "friend." Neither I nor anyone else in the strike had ever seen him before the bust.

My charges were felonies—riot, incitement to riot, and trespass. I was accused of "urging others to commit a felony—to wit, riot in the first degree." Bail was set at twenty-five hundred dollars. Much worse off were two students who had not been occupying Hamilton but were charged with conspiracy to commit murder, riot, and incitement to riot. It seems that while we were being peacefully and carefully arrested in Hamilton, outside on the campus more than a thousand people had met the second police invasion with a fury of their own.

Angry students built barricades at both entrances to College Walk, using wooden police barriers, garbage cans, and metal loading ramps. Skirmishes broke out as enormous crowds screamed, "COPS MUST GO!" and "UP AGAINST THE WALL, MOTHERFUCKER!" People threw bricks and rocks through President Kirk's office windows; fires broke out in two buildings, Hamilton and Fayerweather. The latter did little damage, just smoke from a burning couch, but a fire on an upstairs floor of Hamilton Hall did destroy the papers of history professor Orest Ranum, who was researching a book on Louis XIV.

The Strike Committee and I always denied any involvement in the Hamilton fire, but the truth, known to only two people, was otherwise. Right before the cops broke in to Hamilton, after a whole night of debating and arguing at an interminable meeting in the lobby about whether to make our stand or not, JJ took me aside and asked, "I want to set a fire upstairs. These motherfuckers have got to fall."

Caught up in "total war" mode, beyond rage, without limits anymore, I replied, "Okay, go ahead," and with those words JJ disappeared up the stairs. I didn't know he was going to target Professor Ranum, but I suspect that he knew exactly whose office he was breaking into. JJ hated hypocritical liberals, and the professor, who had tried to stop the original occupation of Low by saying he was sympathetic to our ends but not our means, had become an opponent of the strike.

JJ and I had crossed over the line of nonviolent protest.

Outside on campus the confrontation with the cops was beginning to die down around 2:00 A.M. At that moment President Kirk, in another one of his classic blunders, ordered the police to clear the campus of everyone, including students. So, for the second time, at around 4:15 A.M., approximately a thousand police stormed the barricades, beating people as they came. What followed was a second and in many ways even more vicious night of violence. The TPF and the plainclothes cops went crazy, beating students with clubs and nightsticks, chasing students around campus, including into dorms, which were supposed to be safe havens. They even pushed a girl through a plate-glass window. Ray Brown Jr., a leader of the Student Afro-American Society, was seriously injured in a beating by six cops. Fifty-six people were injured in total, with a like number arrested. No officers were ever disciplined for their rampage, nor were any supervisors reprimanded. The police riot was never reported outside of the campus and the "underground" alternative press.

After the second Hamilton bust, the university was so polarized that it could not hold its regular commencement on campus. Over four hundred graduates staged a walkout demonstration from Columbia's official ceremony, which for the sake of security was held at the heavily guarded Cathedral of St. John the Divine on Amsterdam Avenue.

Ted Gold, who was graduating, came up with the idea that the walkout should be signaled by the broadcasting of Bob Dylan's "The Times They Are A-Changin'" to transistor radios hidden in the radical graduates' robes. Teddy and a striker who was a radio engineer put a low-power transmitter in my apartment across the street from the cathedral, with an antenna on the building's roof. Had the scheme worked, it would have been a wonderful background song for so dramatic an event, but the puny little transistor radios of the time didn't make enough collective sound for anyone to hear in that enormous cathedral, the third-largest in the world. "Your old road is / Rapidly agin'. / Please get out of the new one / If you can't lend your hand / For the times they are a-changin'."

The Strike Committee conducted our own liberation commencement at College Walk, followed by a rally and picnic at the gym site, where work had been stopped. Fearing the possibility of an additional arrest and on the advice of my attorney, Sue and I went out to Jones Beach. It's a decision I've always regretted. We listened to WKCR for news of the countercommencement on an FM radio we'd brought along with us.

The mood at the liberation commencement was celebratory. One of the speakers was the noted psychologist Erich Fromm, who quoted from Nietzsche: "There are times when anyone who does not lose his mind has no mind to lose."

The former president of Sarah Lawrence College, Harold Taylor, conferred degrees on the strikers:

> By virtue of the authority vested in me by the trustees of the human imagination, derived from the just powers of human nature and the constitution of mankind, I hereby confer upon all of you here present, in addition to all those absent, a degree of beatification through the arts, in order that the arts may flourish everywhere, and the delights of poetry, political action, dance, theater, and the insight of science and the fruits of technology may descend upon you, everywhere on the earth and in the great space surrounding this small planet, for the best use to which it can all be put in the interest of human beings everywhere and to the ultimate benefit of all living things, with the rights and privileges pertaining to our union with nature and our peculiar human condition. . . .
>
> To others who wish to go through life without diplomas or certificates of any kind, I confer upon you a degree of happi-

ness and freedom, and send the best wishes of this outdoor congregation of celebrants of freedom that you will never become addicted to or corrupted by the idea that in order to learn what you need to know, you have to commit yourself to an institution.

May you all learn forever.

SDS rented a fraternity house for the summer on West 114th Street, across the street from Columbia, for our "Liberation School." Hundreds of people came and went at all hours, arguing about Marxism, debating strategy, trying to plan how to broaden the strike and spread the organizing to high-school students, active-duty GIs, and workers. We put out our own weekly newspaper, called *Struggle.*

Throughout May and June, we were intensely aware of the revolutionary events taking place in France. A student strike in Paris had led to a society-wide uprising against the conservative de Gaulle government and, it appeared, amazingly, against the capitalist system itself. Young factory workers joined with students in the streets in one of the most surprising events of our time—a spontaneous revolution in an advanced industrial capitalist country. At one point it seriously looked as if the government of France would fall.

We went wild when we saw on the front page of a newspaper a picture of a French student carrying a sign with just two words: COLUMBIA, PARIS. This was confirmation that we were a recognized and significant part of a worldwide revolution. One day Strike Central received a telegram from the Sorbonne University in Paris: WE'VE OCCUPIED A BUILDING IN YOUR HONOR. WHAT DO WE DO NOW? I don't remember our answer.

That summer we sent Lewis Cole as our representative to an international radical students' conference in England. Lewis was perfect for the job, because he was a tall, rangy intellectual, stunningly articulate, and he already smoked Gauloise cigarettes. When he came back, SDS decided to call an International Assembly of Revolutionary Students at Columbia for mid-September, right before the opening of school. We hoped this conference would reignite the Columbia strike and also infect other schools with our spirit.

When the assembly finally convened, it consisted of a lot of talk coming from very rarefied intellectuals, without any goals or purpose. I more or less ignored the event, concentrating instead on organizing SDS for the confrontation we expected when school reopened.

We didn't have to wait long. During the International Assembly, the new acting president of Columbia, Andrew Cordier—Kirk having been dumped by the trustees over the summer—abolished SDS as a campus organization. We were then denied the rooms we needed for the conference. In response, several hundred people took over a large hall and held our meeting anyway. Abbie Hoffman showed up and gave a hilarious impromptu performance in which he did very skillful yo-yo tricks—with a yo-yo that lit up, no less—while simultaneously spouting revolutionary one-liners to the audience such as, "Manners is a plot by the goyim to keep the Jews out of banking."

The strikers greeted the opening of registration in mid-September with angry confrontations between five hundred of us and several Columbia deans backed by security guards. We demanded that the suspended and expelled students—myself included—be permitted to register.

On its front cover on September 30, 1968, *Newsweek* ran a picture of me in the midst of the crowd at the registration demonstration grappling with a black security guard. The headline read, CONFRONTATION AT COLUMBIA. What really happened in the picture is even more interesting than what appears. Some of the younger guards had been provoked and wanted to have it out with us. I faced off directly against an older guard, Officer Woods, a man with whom I had been friendly for some time. In the midst of the pushing and shoving, he touched my arm lightly and said, "Mark, Mark, what are you doing?" His gentle reproach snapped something inside me, and I immediately put my hands down, pulled back, and told the others on the front line to do likewise. The confrontation stopped.

The strike did die in the fall, despite our attempts to revive it. Students wanted to get back to classes. They weren't paying three thousand dollars in order to go on strike, a good chunk of change in those preinflation times. Even so, it had been one of the longest and strongest student strikes in U.S. history.

The gym in Morningside Park was stopped as of April 23, and the university's ties to IDA were severed that summer. Student and faculty participation in disciplinary matters was institutionalized, for whatever good that did. Most important, thousands of people had become radicalized. That was our biggest victory, the goal SDS had set for itself years before we even knew about IDA and the gym. We wanted to build the movement, and we succeeded.

Though it was far from the first, Columbia became a nationwide symbol for student revolt, making the name SDS into a household word. Right after the Columbia strike, Tom Hayden wrote an article for *Ramparts* magazine in which he used the slogan "Create two, three, many Columbias!" a direct paraphrase of Che's strategy, "Create two, three, many Vietnams." The protests spread even further in the fall of 1968 and through 1969, to San Francisco State, Harvard, Kent State in Ohio, and dozens of other schools. By 1970 hundreds of schools were on strike against the war.

During the strike a black lady from South Carolina wrote to tell me that I had restored the faith in white people's humanity that she had lost after Dr. King's murder. A GI wrote from the siege of Khe Sanh in Vietnam to wish SDS luck and to express support from his fellow soldiers.

To this day I encounter people who tell me the Columbia strike changed their lives: a woman who gave up French literature to study law and work for welfare clients; a male career community organizer who found direction for his life during the strike.

While I was a fugitive, in 1976, I struck up a conversation with a waitress in a coffee shop in midtown Manhattan. She told me she had been in the Columbia strike and that it had changed her life, had given her a new set of priorities. She had become an artist. Perhaps to impress me, she confided that she'd been close friends with Mark Rudd.

For many of us in SDS, especially the Action Faction, Columbia confirmed our original notion that exemplary action leads to mass support and participation. The events of the strike also won over a significant number of our old go-slow adversaries of the Praxis Axis—Ted Gold and David Gilbert among them. Columbia seemed to be a victory for militant vanguard action.

In the SDS National Office and among chapters around the country, Columbia was seen as the culmination of two years of transition "from protest to resistance," as National Secretary Carl Davidson put it. Writing in *New Left Notes,* out of Chicago, he said:

> Since the Columbia Rebellion, SDS has been thrust onto a new plateau as a national political force. The importance of that event in our history should not be underestimated. More than any other event in our political past, Columbia has successfully summed up and expressed the best aspects of the main thrust of our national political efforts in the last two years.

That "main thrust"—overt anti-imperialism, antiracism, confrontation, militancy that would radicalize young people while downplaying institutional reform and student power, peaceful marches, and petitioning—were the lessons of Columbia that would feed a more extreme tendency in SDS. Within a year Columbia would give birth to the "revolutionary" faction known as Weatherman.

PART II

SDS and Weatherman

(1968–1970)

7

National Traveler

fter I was expelled from Columbia, I became a traveling salesman for SDS, visiting chapters and speaking at different colleges around the country. That fall alone I spoke at about seventy-five campuses large and small: UC Santa Barbara, Amherst, Yale, Harvard, Kent State, the University of Pittsburgh, Berkeley, Stanford, Long Beach State College, Los Angeles Community College, and Essex County Community College in New Jersey, to name just a few. At each stop I'd be met at the airport by the local chapter leaders, who were most often men. They'd usually take me to a run-down house or apartment in the area's student ghetto, where we'd talk about their chapter and the issues they had taken up—for instance, some were trying to throw ROTC off the campus, while others wanted to stop the university's expansion into the community; in that period, too, many chapters were demanding an open-admissions policy, which would benefit minorities

and the poor. But in all cases, Vietnam and racism were the underlying concerns, as they were at Columbia.

Many local SDS activists would bemoan the prospect of organizing in what they thought of as right-wing backwaters: "Topeka, Kansas, isn't New York City," they'd tell me. "It's the most conservative campus you'll ever visit." I heard similar complaints in Oneonta, New York; Bellingham, Washington; and Kalamazoo, Michigan. Yet I was seeing firsthand that students were becoming more and more radicalized throughout the country and that organizing was happening everywhere, even in the unlikeliest places. The SDS National Office reported a dozen new chapters in just the first few weeks of September; applications were coming in so fast the membership secretary stopped counting how many chapters there were, let alone individual members.

In a memorable incident at Towson State University in Maryland, Nixon's running mate, Spiro Agnew, then governor of the state, was heckled nonstop as he tried to give his campaign speech. Frustrated, he screamed at the audience, "How many of you sick people are from Students for a Democratic Society?" Nearly two-thirds of the one thousand people in the audience jubilantly stood up and cheered.

I tried to help chapter organizers pick those issues and tactics that would best grow the chapter's membership. Education versus direct actions was the constant tension in our discussions. My advice never varied: With more militancy, I told them, you can reproduce a Columbia in Lafayette, Indiana. I invoked the power of Chairman Mao Tse-tung's mantra, "Dare to struggle, dare to win!"

I often gave a formal talk to the campus, attended by five hundred to a thousand people, sometimes even more. I'd lead off with the Newsreel Collective's hot new film, *Columbia Revolts.* Newsreel members considered themselves "guerrilla filmmakers" because, as participants in the events, they shot from inside. In just two months, they produced a really effective propaganda film that showed life in the communes—meetings, parties, even a wedding—followed by images of police viciously clubbing the peaceful students. The film ended on a high, with scenes from the liberated graduation ceremony—strong, beautiful young people in revolt, marching in the early-summer sun, turning their backs on the ugly military-corporate university.

The lights would then come up, and there I'd be, the star of the movie, standing at the podium *in person* to receive the thundering applause and adulation of the crowd.

In contrast, my talk was usually low-key and direct. I'd first give the reasons for the protests at Columbia—how it wasn't about student power at all but rather about challenging U.S. imperialism and racism. "These are the real issues which affect people's lives, not dorm rules and academic requirements." Then I'd review the events leading up to the revolt: the sit-ins, demonstrations, and dorm organizing. I stressed the need to be bold and to use imaginative action to win over students who were just waiting to act against the war and racism. I'd tell the story of the pie incident, which was always good for a laugh. I'd always end by talking about the issues being pushed by the local SDS chapter.

Though it was mostly a standard stump speech I repeated again and again, I had the kind of energy that gave people the feeling of actual communication taking place. Once, however, a reviewer writing in the University of Michigan student newspaper called me on an obvious verbal trick that I hadn't been aware I was employing. I had developed the habit of pausing to pretend to search for the right word in order to give the impression of spontaneity. My anti-charisma charisma still needed polishing.

In general my speeches were successful, judging by the reactions of the audiences and my radical student hosts. A visitor from the larger movement, I linked people to the national scene and explained what was happening. In addition, that fall I managed to raise tens of thousands of dollars for the National Office of SDS, where staffers were living on $50 a week. A national booking agency was guaranteeing me an incredible $750 net per gig for the talks they arranged, all of which I turned over to SDS.

Inevitably, women would present themselves, or I would find them.

I had broken up with Sue over the summer, and I saw these one-night stands as perks of my minor stardom. I felt no qualms at all about sleeping with whoever was available on the road, even though I had a new girlfriend back in New York. I was a woman junkie. I was fulfilling my longtime sexual fantasy of sleeping with a lot of women. The numbers seemed to prove to me that I was attractive and virile. Sometimes before going to sleep, I'd count the women I'd made love with since I started in high school.

I seemed to attract a funny kind of groupie, though. She was usually serious, studious, and politically involved; the encounter would seem to both of us to be a kind of political event, an encounter with the revolution in bed. Quite often I attracted older women who had just left their marriages or were about to. I was the traveling purveyor of sexual freedom.

My womanizing has been criticized by some feminists who have made me into a symbol of the New Left's mistreatment and oppression of women. I plead guilty: Women were definitely for me a type of object, to be used to build up my already inflated ego. What can I say? I was twenty-one years old at the time, young and opportunistic, living out a standard American male fantasy.

The New Left itself was a male-dominated culture at the time. Men were the theoreticians and orators, the SDS chapter officers; women were the troops and typists. As early as 1966, some women had begun, in all-female caucuses and bitter position papers, to protest their subjugation. They demanded equality as they developed a thoroughgoing critique of male sexism, both in the movement and outside.

We canny SDS male leaders either completely denied the existence of the problem or attempted to undercut the critique by paying lip service to the necessity to fight "male chauvinism and male supremacy" in American society. Meanwhile, it never occurred to us to encourage female leadership by speaking less in meetings or developing other mechanisms for women to be heard. When criticized, we attacked feminists as diverting the struggle from the primary (i.e., more important) issues of racism and imperialism. We created a hierarchy of oppression, in which race and then class both trumped gender. As a consequence, large numbers of women became disgusted with SDS and the other antiwar organizations and quit. They continued meeting together for "consciousness raising," and to work on key women's issues such as abortion rights, employment, day care, and health.

The women who stuck with SDS believed that the fight against the war and racism would lead to an eventual redistribution of power in society that in the long run would benefit women. They still hoped to raise the issue of male supremacy within SDS and the society at large. By 1970 and 1971, however, the struggle within the movement would break out in earnest, toppling male leadership. But by then I was far underground, insulated from the attacks that were my due. It wouldn't be until a few years later that I was confronted by the full fury of women.

When I wasn't on the road, my base was the SDS New York Regional Office (RO), now located in a newly co-oped building at 131 Prince Street, in a declining industrial neighborhood of lower Manhattan later famously known as SoHo. I'm sure lofts in that building now sell in the millions, but at that time we paid two thousand dollars, which a wealthy contributor put up. When our photographic offset printing press wasn't running, we could often hear the great jazz saxophonist Ornette Coleman practicing in the loft above, not that I knew enough at the time to appreciate who he was.

I was a member of the RO collective, which consisted of about a half dozen volunteers whose job was to service the New York–area chapters. We provided organizers, printed flyers, provided study literature, and helped coordinate the thirty or so chapters in the region. Mostly we spread ideas.

I loved being an organizer, putting people in touch with each other, suggesting issues and tactics, encouraging the local campus activists, coordinating events, and maintaining contact with other movement groups. It was pure people work from the time I got up in the morning until I'd drop into bed after a late-night meeting.

That fall one of the biggest issues in New York City was the public-school teachers' strike against decentralization of the school board and community control of the schools. In essence, the teachers' union, led by former socialist Albert Shanker, was on strike against the parents, especially in the black and Latino neighborhoods, who were demanding a say in their children's education. The union vehemently opposed this perceived loss of teacher power, and the fight took on a strong racial content, since the union was predominantly white.

New York City was openly polarized in a way most of us young people found horrifying; this was, after all, the liberal urban North, not the racist, segregated South. I was particularly upset that many of the white teachers and the union leadership were Jews. During that strike I sadly understood that we were experiencing the end of the Jewish-black liberal coalition that had prevailed since the early civil-rights days. It was also the beginning of a hard right turn for mainstream New York City Jews.

Most of the Left, including SDS, supported the parents' demands for community control as part of the larger antiracist battle. SDSers from Columbia and other chapters would ride a subway out to Brownsville, Brooklyn, very early in the morning to join black parents in keeping a junior high school open against the striking teachers. I myself taught in a strike-breaking liberation school in West Harlem. I vowed at that time that I'd never join Shanker's racist union, the American Federation of Teachers (a vow I would break in 1990 when I helped organize the teachers' union at my community college).

But a small faction of New York SDS—mostly the PL caucus from Columbia who had converted en masse from Maoism to a cult form of Trotskyism led by a quirky old economist named Lyndon LaRouche—put out a leaflet and public statement attacking community control as a Ford Foundation conspiracy that was designed to defeat true working-class organization, such as the teachers' union. Albert Shanker used this statement at one of his press conferences to claim that even SDS supported the teachers' strike. To the rest of us, the so-called Labor Committee's position was a betrayal of our fundamental principle of antiracism: They were saying, in SDS's name, that black and Latino people did not have the right to control their own schools.

SDS was now a house divided. After much soul-searching, an assembly of New York regional SDS membership expelled the Labor Committee from the organization. Such a move had never happened since 1962, when the Port Huron Statement put forward the twin principles of nonexclusion on the basis of politics and opposition to anti-Communism. In December the SDS National Council upheld the action.

I joined the fight for the expulsion, which I felt was necessary to keep SDS from being mistakenly seen as supporting racism. Six months later we SDS regulars would use the New York precedent to throw the Progressive Labor Party out, and the organization would split irreparably.

In late October 1968, I received the most ominous letter a young man could receive at the time: a notice from my local draft board informing me I had been reclassified as 1-A, eligible for induction. I was ordered to report for a physical examination.

Monthly draft calls had grown since 1965 to a high of fifty thousand in order to supply the soldiers for the ever-enlarging invasion and occupation of Vietnam and the rest of Southeast Asia. Escalation of the war was the only strategy the United States had, and it was a losing strategy at that. I had been protected up to now by my student deferment, in effect a privilege granted to those who could afford college. But since Columbia had thrown me out in May, I was now fresh meat, the same as anyone else.

My first reaction was terror. Like millions of other men who faced the dilemma of whether to fight in a war they didn't believe in, my options were not good. I could either flee to Canada, openly resist induction and go to jail, or find some kind of new deferment. Instead I picked another route: reverse psychology. I would tell the U.S. Army I wanted in.

By 1968 many GIs had already begun to refuse orders to go to Vietnam. In Vietnam the epidemic of troops avoiding combat and attacking their officers (called "fragging," due to the practice of rolling fragmentation grenades into officer quarters in the middle of the night) had also begun. New York SDS was involved in helping organize a GI coffeehouse outside Fort Dix, New Jersey, one of almost a hundred such projects around the country. These coffeehouses were places where soldiers could decompress away from military control, meet other antiwar GIs and civilians, and eventually organize antiwar publications and protests. By 1970, antiwar organizing within the military would put the army especially on the defensive.

I sought out Andy Stapp, the young organizer of the American Servicemen's Union (ASU). One of the first GIs to resist orders for Vietnam, Andy had been court-martialed, imprisoned, and then discharged. He was now devoting himself to organizing a soldiers' union, similar to a labor union; the ASU put out its own newspaper for GIs, the *Bond.* Andy immediately saw the publicity value of working with me. He gave me information on dozens of GI rebellions, and we discussed the importance of the soldiers' antiwar movement—namely, that it could make the military useless in Vietnam.

My first shot was to send the draft board a formal appeal of my 1-A (ready-to-be-drafted) classification. Not too seriously, I claimed an occupational deferment on the grounds that my occupation, revolutionary, was vital to the national interest. I was just setting the stage. The only reply I received was through the Newark daily newspaper, which quoted the head of my draft board as saying, "Rudd will get out of the draft over my dead body."

Two days before I was to report for my preinduction physical, I held a press conference at the Hotel Diplomat, a Times Square hotel owned by the family of my attorney, Gerry Lefcourt. Several dozen reporters and cameramen attended. I read a lengthy statement saying, in part, "If forced to, I will enter the army; however, I will continue organizing within the army as I have done outside, since my life is committed to the revolutionary movement for freedom, democracy, and peace." I also spoke about the GI movement, citing several cases of rebellion, harassment, and arbitrary denial of human and constitutional rights by the military. I lectured the reporters, "An army of men who think and make their own decisions based on their beliefs in democracy . . . is one of the most dangerous contingencies that faces the rulers of this country."

In order to dramatize how "the power of the small class that controls and exploits our country . . . [would] one day be smashed by the power of the American GIs and American people fighting together for their freedom," I then broke in half an officer's ebony swagger stick that Andy Stapp had given me—he had taken it from a jeep while he was in the army.

A *New York Times* reporter afterward pointed out to me that the stick hadn't snapped in two with a crack; it had just sort of slurped. The night before, I had been nervously testing it, worried that I wouldn't be able to break the damned thing in front of the cameras. Inevitably, I broke it and then used superglue to put it back together, hoping that it would give a passable effect. The show must go on. I don't think many people were fooled.

The truth behind my press-conference bravado was that I was terrified down to my gut that my bluff wouldn't work. I could just see myself in the army, subject to arbitrary military discipline and "justice." It's one thing to organize at Columbia University but another to be insubordinate in basic training at Fort Dix. Army jails are not very nice, I had been told.

The night before my physical, I stayed at my parents' house in Maplewood. I'd heard that eating eggs could increase the levels of albumin in urine, which sometimes could get you a physical deferment. So I slowly ate a huge egg salad I had made with the whites of several dozen hard-boiled eggs. The next morning a hundred people showed up outside the Newark induction center, not only to protest my being drafted but also to support the GIs who were resisting service in Vietnam. Most were from SDS chapters in New York City and New Jersey, including the Newark campus of Rutgers University. Seeing these demonstrators cut away the worst edge of my terror.

Inside the induction center, I began handing out leaflets I'd written about the illegality of the war and about GI resistance and rights. I told the other young men there that if they were drafted, they should consider joining the American Servicemen's Union. The atmosphere of the induction center was very subdued, probably because the other guys were as scared as I was that we would be taken.

I walked up to the first officer I saw and handed him a leaflet and a letter from Michael Klonsky, national secretary of SDS. The letter said I was an admitted Communist and that I would probably cause "havoc, even anarchy in the ranks," especially if left "alone in the barracks with the boys." This didn't seem to impress the officer. As the day wore on, I endured the whole procedure, standing around in my underwear, taking written tests, submitting to a physical examination.

At the end I got to a table where a weary middle-aged doctor asked if I'd ever been treated for a mental disorder. A bell rang in my little brain as I suddenly remembered the four months of useless psychotherapy I'd had in 1966. "Yes, I've seen a psychiatrist," I said. Immediately all the military personnel hovering around started smiling and congratulating each other with relief. They took me into a separate office and told me I had thirty days to get a letter from the psychiatrist, and I'd be given a deferment. My psychotherapy had not been a total waste.

Outside the center, a crowd of reporters was waiting. Asked what it was like inside, I replied, "A normal day at the meat market—they had us up on the meat racks."

The next day I called the psychiatrist, Dr. Liebert. He seemed happy to help the antiwar cause, and my own, by writing a letter to the draft board. Years later, looking through a large file of my Freedom of Information Act papers, I ran across an FBI report that quoted the psychiatrist's letter under the heading "Rudd's problems":

> Chronic depression and a sense of isolation with intermittal [*sic*] suicidal ideation and impulses; high levels of anxiety in relationship to work, dating, sexuality, career and the future generally and difficulty in acceptance of the established authority of institutions.

At first I thought the shrink had laid it on a little thick, but then I realized his description sounded like everybody I knew.

A week or so later, I received in the mail my 1-Y, temporary psychological draft deferment. I felt not the least shred of guilt about using my class privileges, first with the student deferment and then with this one. Everyone should avoid the draft, I felt. If no one answered the draft, there would be no war. The official position of the "draft resistance" movement was, get out of it any way you could. Several thousand young men did take the highly principled stance of refusing to serve. They faced long trials and imprisonment. I rationalized my own position by saying that I would be much more effective fighting the war on the outside rather than from a prison cell.

In 1975, as the war ended, James Fallows, a writer exactly my age, published an article in the *Washington Monthly* arguing that all the middle-class students like himself, an antiwar Harvard student who had gotten a medical deferment, had only transferred the burden of the war onto the poor. This was exactly the channeling function the war planners intended with their deferment policy in the first place: use as troops the poor and minorities, not the middle-class kids.

Fallows argued that the antiwar draft resistance should have followed a strategy of either massive draft refusal, which would have clogged up the courts and the jails, or entry into the military, which would have made the army even more antiwar than it was. Either way the result would have been millions more middle-class families crying out for an end to the war in order to save their sons.

Fallows was right: Draft avoidance was not resistance; it was built into the system. Had I to do it over again, I would not have cooperated with the draft even to the level of accepting a deferment on the grounds that Vietnam was an illegal and immoral war. I would have urged others to do likewise, rather than accept our class-based deferments.

My only participation in the presidential election of 1968—I had turned twenty-one but wouldn't have dreamed of voting—was to speak at an Election Day demonstration called by Southern California SDS in Pershing Square, downtown Los Angeles. In my best rabble-rousing voice, I told the crowd of about 150 that the vote was "a choice between Tweedledee, Tweedledum, and Tweedle Dumber." A significant number of antiwar young people had worked for Eugene McCarthy in the primaries ("Clean for Gene" was their slogan); after the police riots at the Democratic Convention in Chicago and the nomination of the pro-war Hubert Humphrey, most of these kids had no one to support.

The election only confirmed what we radicals had been saying, that in this country there is no real choice; the ruling class controls both parties. Richard Nixon, the Republican who claimed to have a secret plan to end the war, beat Hubert Humphrey, the Democrat who had pledged to keep LBJ's disastrous war going. SDS's alternative was to "vote in the streets," to keep up the antiwar pressure through militant protests.

At a mass meeting held in a Washington, D.C., church basement the night before Nixon's first inauguration, I found myself debating David Dellinger, the avuncular pacifist, "father of the peace movement," and soon-to-be Chicago 8 defendant. I had long respected Dave because of his courage and consistency over many years in the movement—he had been a conscientious objector during World War II, but at that moment I considered him just another liberal pacifist who was unwilling to fight for revolution. I spoke vehemently for blocking the inaugural parade route and attacking Richard Nixon's limousine on the inauguration route. Dave, a partisan of nonviolent though militant civil disobedience, argued for discipline and restraint. We SDSers didn't see what was all that radical about demonstrating peacefully, so we decided to go out on our own.

As it happened, only two thousand out of a much larger counter-inaugural turnout of around ten thousand followed SDS. The rest chose to demonstrate peacefully with the National Mobilization Committee. I ran with my girlfriend and other SDS people from New York. When we got to Pennsylvania Avenue, several people threw rocks and bottles at Nixon's armored limousine, but the police kept us from actually stopping the parade. In a fleeting and chance moment of personal triumph, I managed to throw the finger at Nixon as he glared directly at me out his limousine window, not twenty feet away.

A few demonstrators were arrested at the parade route, and the rest of us "crazies" took off through downtown Washington, stopping traffic, running and shouting in the streets. That day eighty-one people were arrested. But it was as if the counterinaugural demonstration had not occurred at all—there wasn't even a word of it in the papers.

A few days later, back in New York, I got a call from Walter, an acquaintance in the black high-school movement in the city, who said he had watched me during the demonstration and that I had "just been running with the girl" and done nothing violent at all. I had seen him and a partner throw a garbage can through a bank window. He said he now knew I was a phony and couldn't be relied on in the coming revolutionary struggle.

I was devastated . . . because his criticism was true. Since I was a little kid, I'd been (and still am) afraid of violence. I was always ashamed of not standing up to bullies, even when they directly challenged me. In my family, violence was for the goyim or the *trombeniks* (hoodlums). I knew no one who hit another person. I did not play football. I had never lost my fear of violence. But as a member of the cult of Che Guevara, I had evolved a belief in the necessity for violence in order to end the war and also to make revolution. I must have repeated a hundred times in meetings and speeches, "The ruling class will never give over power peacefully." At the time my friends and I thought this line was rational and logical: No contemporary revolution worth anything had been accomplished without violence. We mouthed Mao's famous slogan, "Political power grows out of the barrel of a gun."

Seeking to emulate the revolutionaries we admired in Cuba, China, and especially Vietnam, we convinced ourselves that violence would be successful in this country. We saw the black-power movement, led by the Panthers, already fighting a revolutionary war from within the United States. In our heroic fantasy, eventually the military would disintegrate internally, and the revolutionary army—led by us, of course—would be built from its defectors.

But as I postured and gave speeches on the necessity for violence, I was terrified. I and most of my comrades were middle-class white kids trying to prove ourselves in the world of black, working-class, and international revolution. No wonder that call from Walter was so disturbing, and remained so for years after: I had been found out.

Within a week of the counterinaugural demonstration, I was constantly achy, had lost all my strength, and was sleeping a lot. I was relieved to discover that I had nothing more serious than mononucleosis, which at the time was known as the "kissing disease."

The cure was to go to bed for six weeks. This was a welcome timeout—a mini-breakdown—from my constant running around, travel, meetings, public appearance, more meetings, and talk of revolutionary violence. I loved just lying in bed, reading and listening over and over to the Rolling Stones' new album, *Beggars Banquet,* which celebrated the Street-Fighting Man.

My parents would periodically deliver chicken soup or spaghetti and meatballs to my apartment on West 110th Street, to help their revolutionary son recover.

By late March I was back on my feet—and on the road again, this time for a speaking tour of colleges in the South, where I had never been. SDS was organizing in the bastion of racism.

Since Nixon took office in January, his vitriolic new attorney general, John Mitchell, had been going around the country railing about the need to defend campuses from the anarchic violence of coercive radical students and to clean the streets of black revolutionary troublemakers. In predawn raids on the morning of April 2, 1969, thirteen members and leaders of the Black Panther Party were arrested on charges of conspiring to commit murder and arson, and eight more were sought. The sweeping nature of the Panther 21 arrests, plus the absurdity of the charges—including conspiracy to blow up New York City police stations, the New Haven Railroad tracks, department stores, and for good measure the New York Botanical Garden up in the Bronx, caused us in SDS and on the Left to believe that the conspiracy was the government's, not the Panthers'. In time we were proved right.

New York regional SDS and other movement organizations, both black and white, set to work defending the Panther 21. The morning after the arrests, I was with several hundred people who showed up in front of the criminal courthouse to protest the arraignments. The next day four hundred of us demonstrated at the court, then marched to the Women's House of Detention in Greenwich Village, where the female Panthers were being held. A week later, after intensive publicity and organizing efforts led especially by SDS, two thousand people, including 250 uniformed Panthers, rallied outside the courthouse at 100 Centre Street, site of "the Tombs," while bail hearings were going on inside.

The judge would not bring the defendants from their various jails for the hearing and refused to lower the bail below a hundred thousand dollars per person. He also cited the three Panther attorneys—

William Kunstler, Arthur Turco, and Gerald Lefcourt, my own lawyer—for contempt when they argued that the exorbitant bail constituted preventive detention and was part of a nationwide strategy to smash the Black Panther Party.

I was outside the courthouse that day, and when we heard what was happening in the courtroom, the crowd went wild. We set off for Wall Street, the symbolic center of U.S. capitalism, smashing windows in stores and limousines along the way. My Freedom of Information Act papers include a report from an agent assigned to tail me: "Subject was observed leading [the march], setting trash cans on fire." I had no idea I was under surveillance, but had I known, I probably would have done the same thing, because I was in such a rage about what was happening to the Panthers.

Back at the Tombs, the police were beating up Abbie Hoffman in a phone booth in the lobby, where he had stopped to make a call on the way to one of his own court appearances. Abbie was no stranger to police beatings.

It took more than two years—in one of the longest criminal trials in American history—for the Panther defendants to be cleared by a jury who felt it was inherently wrong for the government to infiltrate the group and set up the acts of violence alleged in the conspiracy. The defense effort required the full-time commitment of dozens of volunteers, including many attorneys such as my friend Gerry Lefcourt, from the National Lawyers Guild, a nationwide bar association of activist leftist lawyers. Defending the Panthers also cost hundreds of thousands of dollars, all raised through donations from both rich and poor. Despite its humiliation in court, the government was quite successful in its principal objective: After the Panther 21 case, the Black Panther Party was never again effective in New York.

On May 5, 1969, the *New York Times* ran a feature article on SDS that judiciously examined our successes and "potentially grave crises." Embedded in the middle of the article was a paragraph that read, "Perhaps the single most pervasive SDS problem, and one that appears likely to grow under the Nixon administration, is police surveillance and infiltration." Ironically, to illustrate its point, the article used the false story the paper had printed the year before about the undercover cop who had arrested me during the second Hamilton occupation at Columbia as being my "trusted bodyguard." I had never seen or heard of the guy.

The real surveillance and infiltration, however, were indeed enormous. The first time I'm referred to in an FBI file was in January 1968, long before I became known to the press. An FOIA document lists me as one of twenty students traveling to Cuba that month. By May 1968, during the Columbia strike, FBI analysts were writing long profiles based on newspaper articles and agent reports. Much of what they had to say was pure fantasy, as in an ideologically biased story from a Caracas, Venezuela, newspaper claiming that I had I met with Fidel Castro and "other hotheads" in Cuba to plan "modern revolutionary struggle." The article also said, "It was specifically proposed at the meeting that sabotage, fires and looting in American cities be greatly intensified." It's likely that this story was planted in the Caracas paper by U.S. disinformation operatives, a practice later exposed by former CIA agent Philip Agee. But it then turned up in FBI files as the truth, a fine example of confusion between propaganda operations and actual surveillance.

The *Times* article, without citing specific sources, but with some claim to authority, went on to say:

> But across the country, federal, state and local investigators keep scrupulous track of SDS activities. The Federal Bureau of Investigation maintains lengthy dossiers on all of its important members and has undercover agents and informers inside almost every chapter.

We'd all been aware of the surveillance for quite some time. Back in 1967, Jeff Jones, who was then the New York regional organizer, had surprised a man who was attempting to break in to the SDS Regional Office, which at that time was located in an old office building on Union Square. The guy told Jeff he was a private investigator with an office in the same building. He said he thought it was his own office and that he'd forgotten the keys. What could Jeff do, call the police? We were in a state of war, and these things had to be expected.

In mid-May I was in Detroit, after a Revolutionary Youth Movement (RYM) conference, wondering how to get home to New York City. The whole contingent from New York, including several people from Columbia SDS, was driving straight back. I wanted to stop upstate, in Ithaca, to spread the word about the RYM to the very large and influential Cornell SDS chapter. My roommate, Peter Clapp, and I borrowed a car—a dark blue 1964 Chevy Impala convertible—from a Columbia SDS woman. She in turn had borrowed it from her boyfriend, a wealthy law student. We promised to return it in a few days.

Leaving Detroit on a bright spring morning, we pulled out a road map and saw that the shortest distance to Ithaca was straight through Ontario, Canada, to Niagara Falls. We thought we'd stop and see the falls—as Peter had never been there—and then cruise on down to Ithaca that evening. As we drove, we soaked up the spring sun with the Impala's top down. There were shades of my not-too-distant childhood, when my father had owned a series of Ford convertibles, including the infamous 1959 Skyliner, with the hardtop that folded up and went into the trunk at the touch of a button, the very car that the Americans had taken to Red Square in Moscow to show capitalism's superiority.

We cruised on through the gently rolling fields of Ontario, happy as the day was long despite the war and all the other manifestations of capitalism we had vowed to overthrow. We agreed that it was such a relief not to have to worry about the pig harassment or surveillance. We pulled in to the Rainbow Bridge at Niagara Falls to cross back over to the U.S. side. Maybe it was the flashiness of the Impala, or maybe we fit the U.S. Customs Service's profile of drug smugglers or draft dodgers, but the inspector on the U.S. side told us to get out of the car and submit to a search.

We both emptied our pockets onto a counter, quite unconcerned since we were legal and had no contraband of any sort. A miniature version of the Little Red Book, *Quotations from Chairman Mao Tse-tung,* tumbled out of Peter's pocket. He had attended high school in Taiwan, when his father was stationed there in the State Department. China's anti-Soviet revisionism was in vogue with our clique, and he liked Chairman Mao.

"What have we here?" exclaimed the wary inspector. "Let's see who you are."

While the customs agents wired our names to Washington to see if we were on any wanted lists, Peter and I were taken out to watch five men go through the whole car, including our bags in the trunk. I still wasn't too concerned, believing that there was nothing they could get us on. Suddenly one of the agents held up in triumph an empty hash pipe he'd found in the back of the glove compartment. Great! First a Little Red Book and now a hash pipe with a tiny amount of residue in it.

I suppose disaster doesn't always slap you in the face. This time I was so relaxed, still basking in the afterglow of the glorious spring day, that I wouldn't admit to myself we were in a predicament. We'd surely scrape through.

They put us in a room and told us to wait. Through a window we could see various officials scurrying about, conferring with each other. At one point an agent came in to tell us that our names had come up on a computer watch list. My sun-drenched confidence was still intact, though.

After about an hour, we were told to go out to our car, where another search was in progress. Opening the trunk for at least the second time, an agent jubilantly discovered a plastic bag with about two ounces of what they claimed was marijuana right on top of our sleeping bags.

If the pot had been there all along, the agents would have found it the first time they opened the trunk. Also, we would have smoked it long before. Loose as we were, we certainly wouldn't have chosen to go through Canada from Detroit, risking a customs search on reentering this country. The customs agents had planted it in the car.

There was one bizarre hitch to this setup: That same day the Supreme Court had ruled in a case of Tim Leary, the guru of LSD, that federal laws against the importation of marijuana were invalid because it could not be proven that the pot had been produced outside the country. Theoretically, it could have been grown here, then taken out, then brought back. The federal customs agents easily solved the problem, however: They turned us over to the Niagara Falls police on New York State charges of possession of marijuana.

Before I was taken off to the county jail, I called a local contact, the head of the Buffalo chapter of Youth Against War and Fascism (YAWF), a small anti-imperialist group with which SDS in New York City had good relations. The guy called a local attorney for us and also came out to the customs station to see if he could be of any help. I gave him my address book, which doubled as a notebook, because I didn't want it to fall into the hands of the FBI.

Finally, while we were being processed at the county jail, in Lockport, New York, the full shock hit me. That morning we had been almost carefree, feeling the freedom of wind and sun on our faces in the borrowed blue Impala convertible. Now the dingy, cold tile cellblock walls forced the reality to sink home: We were prisoners in America, set up on phony dope charges.

Two days later a New York stockbroker I knew put up three shares of IBM as collateral, and we were released on bond. He must have lost his stock when I later failed to appear for trial. Meanwhile, our friends at Columbia SDS organized a fund-raising dance to pay for legal expenses; the slogan, of course, was "Free Rudd and Clapp, or Niagara Falls!"

Years later, when I got copies of some of my FBI files, a complete photocopy of my address book turned up among the hundreds of pages. My contact from YAWF had been an informer and had turned over a copy of the book to the FBI. I had figured that out within a few months of the bust. An SDSer from Newark, New Jersey, came to Chicago to confess to me that he'd been an informer for the House Un-American Activities Committee. He had been shown a page from my address book with his name on it when they recruited him.

The memo accompanying the copy of my notebook from the FBI files is headed "Activities of US Black Militants" and dated June 3, 1969. Presumably, there was a separate intelligence section just to keep track of the blacks, and a hookup with the mostly white SDS was something the government was watching out for.

The memo falsely claims that the notebook was confiscated along with all the contents of my car. It also says that the local FBI and Army Intelligence were provided with copies. The last paragraph of the memo is significant:

> Since there has been a now-and-then tie-in between the SDS and the Black militant groups, we thought this document might apply to this case. In any event, it seems likely that Office of Security might like to peruse the lists of Rudd's friends and acquaintances.

Along with hundreds of names, addresses, and phone numbers, the notebook was full of my own notations about speaking engagements, income, contacts at high schools and colleges, plus notes concerning my contacts with SNCC, the Black Panther Party, New York Newsreel Collective, National Welfare Rights Organization, the United Black Brothers at the Ford plant in Mahwah, New Jersey, among other groups. Included also were outlines for articles I was writing for *New Left Notes,* the SDS national publication, as well as meeting notes.

In one entry I summarized a discussion of our organization's security predicament at an SDS National Interim Committee meeting held the previous month in Chicago.

National repression:

—Security: personal letters, contacts, address books, records

—2nd and 3rd level leaders.

This latter point refers to the need to develop backup leadership in case any of the top people were busted, jailed, or worse. We expected repression.

I have no way of knowing how useful the copied notebook was to the government, but it proved a gold mine for me decades later in trying to reconstruct my activities for the early months of 1969. My life, looked at from this perspective, seems like a whirlwind of meetings with SDS chapter leaders, traveling for speaking engagements, networking with other organizations, local and national strategy meetings. That was an organizer's notebook the government stole, documenting a particularly intense period.

8

SDS Split

SDS was always a loose confederation of local chapters, more of a conglomeration than a single unified organization. Each chapter could decide what issues to work on and what positions to espouse. There was no constitution to guide it, only the Port Huron Statement from 1962, which provided an orienting spirit of "participatory democracy," whatever that was. In 1969 it wasn't likely that too many of the active members, though, had even read our founding document.

Four times a year, SDS held National Council meetings, which nominally represented the chapters, but chapters weren't obliged to follow national decisions. More often than not, the National Council or the National Convention (once a year in the summer) would mandate a program, and then little or nothing would result from it. Such had been the fate of the campaign called "Ten Days to Shake the Empire" in April 1968, and the "No Class Today, No Ruling Class Tomorrow," national student strike mandated for Election Day, November 1968, both of which almost completely fizzled.

Part of the reason that SDS was so loosely structured lay in the enormous differences between schools—elite private colleges, state universities, community colleges, local high schools. In addition, the varying political geography of the country made some regions, such as the East and West coasts and a northern midwestern tier, more radical than others. Mostly, though, the strong anarchistic strain present in almost all the SDS chapters and members made for the rejection of political "lines" and of discipline to a single center.

But there was one serious exception: Starting around 1965 or 1966, the Maoist Progressive Labor Party had begun infiltrating SDS wherever it could with its members or "cadres." At times they controlled several important active chapters, such as Harvard, City College in New York, and San Francisco State College. Their goal was first to push their political line, which varied according to the whims of PL's central committee, and secondly to recruit members out of SDS for the party. They were an external, disciplined ingredient in our ultrademocratic anarchist soup.

These weren't bad people by any means. Most were from the identical middle-class background as many of the regular SDSers. The biggest difference was one of temperament, but that difference was important. PL people had a sure, tight, almost missionary state of mind that made them certain they held the "correct revolutionary line."

PLers didn't smoke dope or wear their hair long. That would turn off the workers, who, according to the PL line, were serious proletarians and wanted nothing to do with the counterculture. (They never noticed that young workers themselves were starting to let their hair grow and smoke pot.) Adherents of the PL line believed that students were bourgeois, privileged, middle-class, and could only be effective if we allied with the working class.

In true religious fundamentalist fashion, the Progressive Labor Party had a central article of faith: The *workers will make the revolu-*

tion. Why? Because Karl Marx had said so back in 1848. "Objectively," they would argue, "the central contradiction is between the capitalists and the workers." ("Objectively" meant that it didn't matter what people perceived or thought, that the truth is constant and unavoidable.)

For PL, black people fighting for their rights against racism were merely the struggle of a sector of the working class, not victims of white racism and oppression seeking freedom. Racism, in fact, was merely a trick the ruling class used to divide the workers; it had nothing to do with white-supremacist culture or real economic privilege that accrues to white people in this society. To PL, the Black Panther Party was both too nationalist, since it put forward separate demands for black people, and at the same time too socialist, because there could only be one true vanguard party of the workers, and that, of course, was the Progressive Labor Party itself. The rest of us thought this view of black revolutionaries was racist.

In a related ideology, PL members believed that all Third World nationalism, such as the struggle of the Vietnamese against U.S. aggression, was a diversion from the real class struggle. Thus, Ho Chi Minh, the Vietnamese leader, was a sellout because he was a nationalist and also because he was willing to negotiate with the American aggressors (the Paris Peace Talks had started in 1968). We SDSers used to quip that PL would fight to the last drop of Vietnamese blood.

The most pernicious effect of PL was that the SDS regulars, myself included, became convinced that we needed a well-worked-out revolutionary theory—and dogma—of our own that would counter theirs. We found one, coming from Cuba and SDS's experience at Columbia University.

In my travels around the campuses that year, I discovered pockets of SDS activists who felt as I did, that bold action—propaganda of the deed—would cause more people to become radicalized and join us. This seemed to be the main lesson from Columbia, a direct translation of Che Guevara's strategy of revolt in Latin America to the imperialist homeland.

We would become urban guerrillas, rejecting the go-slow approach of the rest of the Left, just as Che and Fidel had had to reject the Cuban Communist Party's conservatism by beginning guerrilla warfare in Cuba. Our bible was a slim paperback, *Revolution in the Revolution?,* by Régis Debray, a young French intellectual who had been imprisoned in Bolivia for spending time in the jungle with Che. The first time I saw a copy was in the summer of 1967 when I stopped at the SDS National Office in Chicago and a staffer handed me the book, saying, "This is our future."

At almost all of the chapters I visited from the fall of 1968 to the summer of 1969, SDS members wanted to know more about the Action Faction–versus–Praxis Axis split that had transpired at Columbia before the strike. I always replied that only after the go-slow talkers and intellectuals were thrown out of leadership was the chapter able to take the actions that provoked the victory at Columbia. I was fond of saying at the time, that "'organizing' is another word for going slow." And I wanted to go *fast.* There wasn't a minute to lose.

Similar splits had taken place at the University of Michigan chapter, led by Bill Ayers and Jim Mellen; in Kent State, Ohio, led by nineteen-year-old Terry Robbins as regional traveler and a local cadre of committed-to-the-core student organizers; and at the University of Chicago, under the leadership of my old friend from Columbia SDS, Howie Machtinger, now a sociology graduate student. All these chapters had developed very militant struggles around recruiting, institutional racism, or the universities' ties to military research—the same issues that had spurred the uprising at Columbia.

By late 1968 I joined with a dozen or so people from New York, Michigan, and Ohio and the National Office in Chicago to formulate a response to Progressive Labor. Calling ourselves the National Collective, we presented a series of "strategy" papers to SDS national meetings starting at the December 1968 National Council in Ann Arbor, Michigan. Our answer to PL's Worker-Student Alliance was something we called the Revolutionary Youth Movement (RYM).

As our position developed over the course of the spring, we argued that any revolution in the United States would take place within the context of the international revolution against imperialism. This seemed self-evident from what was happening in the world. The black struggle for liberation would free white people from our own racism, as we acted in solidarity with revolutionary groups such as the Black Panther Party. We proposed that SDS students move off campus to organize young people in high schools, at jobs, on the streets, and in the military for the emerging revolution. We would need organizational discipline—here the word "cadre" was used—to develop revolutionary action, including "armed struggle as the only road to revolution."

The National Collective was a curious phenomenon in its own right. We merged two distinct factions based mostly on our sworn opposition to PL. The first was grouped around SDS national secretary Michael Klonsky and a political sidekick of his, Noel Ignatin, a Chicago auto worker and Marxist theorist who had come out of the Communist Party. With a classical Marxist worker orientation, this faction would constitute the leadership of RYM II, when it eventually split off after the June 1969 convention. Mike Klonsky went on to found a Maoist party that became just as dogmatic as PL around the working class as the agent of revolution.

The other faction, which included me and Karen Ashley from New York; Bill Ayers, Terry Robbins, and Jim Mellen from the Ohio-Michigan region; Bernardine Dohrn, SDS's interorganizational secretary; John Jacobs (JJ), Steve Tappis, Jerry Long, and Howie Machtinger from Chicago; and Jeff Jones, who was then living in California, would soon form Weatherman.

Through the early part of 1969, the National Collective met together roughly every month, but at one meeting, out of the blue, Mike Klonsky made a speech concerning the need for a disciplined revolutionary political party; he was talking about classical Marxism-Leninism. To prove some point, he repeated several times, "Stalin is the cutting edge." Howie Machtinger and I shot glances at each other, as if to say, *Oh, yeah?* This was the beginning of the break with Klonsky, whose adulation of Joseph Stalin made no sense to us. In one of our apartments, we put up a poster of Stalin with a text balloon over his head that read, "Klonsky is the cutting edge."

Soon our faction of the National Collective began meeting periodically without Klonsky, Ignatin, and their allies. In our discussions, JJ and Jim Mellen took on the role of theoretical leaders. Both had been in the now-defunct May 2 Movement, which had been a PL anti-imperialist front group back in 1965 and 1966, before PL moved over to the worker-as-agent-of-revolution line. Both JJ and Jim brought an extreme internationalist perspective, stressing the primacy of Third World revolutions. JJ in addition pushed armed struggle in the Third World as a useful model for us to follow here. We studied the successful revolutionary war in Cuba and the then-current urban guerrilla war led by the Tupamaros in Uruguay. JJ would bring up some arcane question for the group to discuss, such as whether the revolution would happen in one stage or in two. We would debate the point, and then he would go off by himself and write for a few days.

All this discussion was leading up to a strategy paper to be presented at the mid-June SDS National Convention. The paper still didn't have a title on the night that it—all twelve thousand words—was to be sent to the printer in order to be included in the *New Left Notes* convention issue. Several of us in the National Collective were sitting around in the apartment of some Chicago SDS people, brainstorming on the name, and Terry Robbins, who had memorized dozens of Bob Dylan songs, blurted out the line "You don't need a weather man / To know which way the wind blows" from "Subterranean Homesick Blues."

That was it! Just the perfect sort of countercultural dig at the super-straight PLers, and one that made an explicit statement, too: You don't need ancient dogmas to understand the reality around us. Within days the strategy paper became known as "the Weatherman paper," and our faction of SDS became "the Weathermen."

Despite its pop title, the Weatherman paper was absurdly long and melodramatically serious. In it we claimed to know exactly "what is to be done" about building the new American revolution. We had answers for everything plaguing the movement, argued with a self-contained pseudo logic. Its language was much more of the Old Left than the New. Carl Oglesby, former president of SDS, wrote, "Any close reading of the RYM's Weatherman statement will drive you blind."

Gone permanently was the sense of experimentation and openness of the early SDS. If it was going to be war between Marxist factions, we would not shrink from the battle of correct words and ideas. At the time this didn't bother me; I was proud that we were moving from reform to revolution; to make a revolution, you need an analysis and a plan.

The paper led off with a phrase that became the official Weatherman explanation for everything else we would do and say in the next year—actually, in the next seven years: "The main struggle going on in the world today is between U.S. imperialism and the national liberation struggles against it."

Basically, we argued the following: The United States is rich because of a world empire that channels wealth to this country. The revolt now taking place against the empire (e.g., Vietnam) will cause the overextension of U.S. military forces (Che's slogan "Create two, three, many Vietnams"). Internally, the country is undergoing social crisis, including the revolt of black people, who have been an internal colony for hundreds of years. Since black revolutionaries are already engaging in armed struggle, whites should support them ("Share the cost").

According to Weatherman, most white workers have privileges from this imperialist system—that is, relative wealth and status above nonwhite people—and would not oppose the system because of these privileges. However, young people, including young white workers, are experiencing the crises of the system and are rejecting its control (the "youth movement"). We revolutionaries should be organizing in high schools, in the military, on street corners, linking up issues and teaching that "imperialism is always the issue." To build this movement, we will need "revolutionary collectives" and, in the end, a clandestine political party to carry out the inevitable revolutionary war.

Going into the National Convention we had moved from a loose organization of more or less effective campus chapters all the way to believing we needed a highly centralized, clandestine revolutionary party. We had brought the New Left full circle back to the strategy and organization of the Old Left. And we would march proudly into the future at the head of the revolutionary youth column, overcoming all obstacles.

The last SDS National Convention opened June 18, 1969, in the ancient, sooty Chicago Coliseum, on South Wabash, just down the street from Chicago's police headquarters. The National Office (NO) had tried to arrange to hold it at various college campuses where we had strong chapters, but either administrators' hostility or fear, or threats from the FBI had frozen us out. So all fifteen hundred delegates assembled at this cavernous, gloomy roller rink.

After the delegates had been registered and searched for drugs, weapons, and tape recorders and a few recognizable police agents had been ejected (there were dozens more of all varieties outside snapping pictures), the first day of the convention was taken up with procedural questions. In these votes we in the NO faction were surprised that PL had much more strength than we expected. In fact, they won several votes simply by gaining the support of independents who voted solely on the merits of the issues, not on factional lines. This meant that we, the Revolutionary Youth Movement leadership, didn't represent a majority of the delegates.

We did have considerable support, though, located primarily among the Action Faction–dominated chapters. These included most of the delegation from New York, with a particularly heavy turnout from Columbia, and the delegations from Ohio-Michigan, the University of Chicago, and the Northwest, the last of these led by Roger Lippman and Clayton Van Lydegraf, an old Communist-turned-Maoist who soon would become our de facto guru on account of several papers he had written concerning the need for armed struggle.

Many of the RYM delegates were younger students from midwestern colleges who had never been to a National Convention before but were enthusiastic about their role as partisans against the evil PLers. During one bitter debate over some small bureaucratic matter such as allowing the press into the proceedings, a group of about fifty RYMers from Michigan and Ohio, jumping up on their chairs, wildly waved Little Red Books over their heads and chanted, "Mao, Mao, Mao Tse-tung! Dare to struggle, dare to win!" Then they collapsed, laughing at their own little guerrilla theater parodying PL. The PL people, never known for their sense of humor, just scowled back.

In addition to our troops, there were members and even whole chapters who claimed independence and allied themselves with the RYM/National Office leadership solely because they hated PL's dead rhetoric and manipulative organizing style. Some of these people felt similarly about the National Collective's increasingly dead rhetoric and manipulative style yet could be counted on to vote against PL when the showdown came. We shared the independent tradition of SDS, not members of an external grouping.

The opening procedural votes were so close and swung back and forth so much that on the second day, I later learned, the National Office staff doing the counting began fooling with the numbers, adding or subtracting votes as needed to produce anti-PL results. So much for the democracy part of Students for a Democratic Society. Ends and means had long ago become blurred.

The NO staff, which had set the agenda for the convention, had scheduled a show of solidarity with RYM/NO for the second evening, in the form of speeches by representatives of black and brown allies of SDS—the Puerto Rican Young Lords Organization, the Chicano Brown Berets, and, finally, the Chicago Black Panthers. All the speakers attacked "those who say all nationalism is reactionary," a clear reference to PL. The Chicago Panthers' minister of information,

Chaka Walls, lit into Progressive Labor's claims to be the vanguard party. "These armchair Marxists haven't even shot rubber bands," he said, while the Panthers, the true vanguard, were out shedding blood.

Then, without any lead-in or warning, Walls started talking about women in the movement, using the term "pussy power." The crowd was stunned at first, silent, but then the ever-opportunistic PL faction began chanting, "Fight male chauvinism, fight male chauvinism!" Pandemonium broke out in the hall. Walls apparently didn't understand the ideological issues behind the chants, because he added, "We've got some puritans in the crowd. . . . Superman was a punk because he never even tried to fuck Lois Lane."

That did it. The PL chanting increased to a roar. Our people were screaming, too, telling the PLers to shut up. The sound was deafening, like the cheering at a championship basketball game, only everyone on both sides was screaming for murder. I didn't know what to do, nor did anyone else. Within minutes another Panther leader, Jewel Cook, took the microphone and again attacked PL. This cheered us up and shut up the PLers. But then Cook began defending Chaka Walls, saying he, too, was for pussy power and that "the brother was only trying to say to you sisters that you have a strategic position in the revolution: prone."

My heart sank. That stupid joke, which had been around for years, set the whole PL section chanting again, "Fight male chauvinism, fight male chauvinism!" Only this time the roar was even greater.

The Panthers had sunk us. What was supposed to be a show of solidarity from the black vanguard, already anointed as such by SDS in a vote at March's National Council meeting, had turned into a rout for both them and us.

The next day Jewel Cook came back to address the convention with an ultimatum jointly endorsed by the Panthers, the Brown Berets, and the Young Lords: SDS would have to kick out PL, which denied the right of self-determination to oppressed people, or else cease to be considered revolutionary. "Students for a Democratic Society will be known by the company they keep," he warned the auditorium.

The PLers started screaming, "Smash red-baiting, smash red-baiting!" referring to the old anti-Communism that SDS had repudiated long ago—themselves being the Communists who were being baited—while the RYMers responded with the Panther battle cry, "Power to the people! Power to the people!" The Panthers stomped off the stage.

The PL leader, Jeff Gordon, then took the microphone and said, slowly, "PL . . . will not . . . be intimidated out of SDS." More screaming.

Mike Klonsky, Bernardine, Terry, and I conferred at the back of the stage. We didn't know what to do; this ultimatum from the "vanguard" couldn't be ignored. On the other hand, we were reluctant to split the organization, no matter how much we despised PL.

I went to the podium and proposed an hour recess to let things cool down. Even before I was done, Bernardine grabbed the microphone from me, having broken away from Klonsky, who was trying to restrain her, and shouted, "We're going to have to decide whether we can continue to stay in the same organization with people who deny the right of self-determination to the oppressed! Anyone who wants to talk about that, follow me into the next room." I paused for a second, looked out into the massive crowd, then followed her out of the big hall, into the smaller auditorium.

About two hundred people came as well, and the number eventually grew to around five hundred. We debated the pros and cons of splitting. We couldn't be in the same organization with people who were objectively racist and sexist, yet most didn't want to run the risk of destroying SDS over a split.

That night I met with the National Collective, and we still didn't know what to do. The best we could come up with was to meet with the regional caucuses. I was assigned to meet separately with New York and the Northwest. Enervated, I stayed up all night listening to people's grievances against PL, how they had messed up chapter organizing, how they had undercut support for the Panthers and the black struggle in general. At both meetings people demanded overwhelmingly to expel PL. I reported this sentiment back to the National Collective in the early morning. Other regions had also demanded we throw PL out. Thus, through a democratic deliberation process of sorts, it was decided we would make a split.

While PL took over the plenary session in the main hall, our rump caucus met in the arena next door for our final debate. Bernardine made a long, brilliant speech detailing the history of SDS and how PL had always run counter to our principles. She proposed that we march into the main arena and tell PL that they were no longer part of SDS. She ended, "*We* are SDS."

A vote was taken: With about six hundred people present, Bernardine's proposal won, roughly 5–1.

So we did it. The leadership marched into the main arena, taking up positions guarding the stage. Our people formed a ring around the outer aisles of the giant room. Bernardine took the microphone, and the crowd was absolutely silent. I stood right next to the podium with my arms crossed, looking out defiantly at the mostly PL crowd in the seats. For about twenty minutes, Bernardine recounted all of PL's crimes. At last she paused. "SDS can no longer live with people who are objectively racist, anti-Communist, and reactionary. Progressive Labor Party members, people in the Worker-Student Alliance, and all others who do not accept our principles . . . are no longer members of SDS."

At first there was silence, followed by awkward nervous laughter, then cries of "Shame! Shame!" Bernardine screamed into the microphone, "Long live the victory of people's war!" This was the Chinese Communist Party's battle cry; the rest of us began chanting, "Power to the people! Power to the people!" and "Ho! Ho! Ho Chi Minh!" as we marched out into the Chicago night.

The long-feared split had occurred without any full debate by the whole organization, without any vote taken. It was a fait accompli, a coup of sorts, presented by the RYM faction. The next day, while the PL faction convention met back at the Coliseum, about eight hundred to a thousand delegates of what we called "the real SDS" met at a Congregationalist church near the National Office on Chicago's West Side. We had secured the NO the night before, ready for an attack by PL that never came, to seize the mailing lists, books, and print shop.

The rump meeting itself was something of an anticlimax. We passed a ridiculous statement of "principles," hastily put together by our collective. Somebody threw in among the points of agreement solidarity with the People's Republic of Korea and of Albania, both of which we knew little about; we didn't want to miss any good revolutions we should be supporting. We passed two action resolutions, one supporting a Panther United Front Against Fascism conference for Oakland the next month, and the second calling for a fall National Action against imperialism and the war, to correspond with the opening of the Chicago 8 trial in late September.

At the end of the meeting, we held elections for national officers. My collective nominated me for national secretary, Jeff Jones for interorganizational secretary, and Bill Ayers for educational secretary. Now that PL was no longer in the picture, the Klonsky faction and those with a more traditional Marxist "worker" orientation split off from the National Collective, running a slate against us. My "campaign" speech didn't once mention my politics or my history. We had decided not to rehash all the issues of the Weatherman paper and the split during the election. I just said that the movement needed symbols and the media had seized on me as a symbol of SDS. It was a pathetic campaign speech, but I won by a big margin, as did Bill and Jeff.

I was now national secretary of the largest radical antiwar organization in the country, hopeful that the split had cleared the way for SDS to play a much bigger role in the developing revolution. What I didn't realize at the time was that the split marked the beginning of the end for SDS.

9

Bring the War Home!

As the new national secretary of Students for a Democratic Society and a leader of the Weatherman faction, I threw myself into organizing for the National Action to be held in Chicago on October 8 to 11, coinciding roughly with the opening of the Chicago 8 conspiracy trial. In the name of SDS, an organization of almost four hundred chapters and probably a hundred thousand activists, our tiny clique of about ten people, now known as the Weather Bureau—with a base at only a handful of colleges—issued a call for the first gathering of the revolutionary youth of the United States to join us for three days of violence to match the violence of the United States in Vietnam. Even our erstwhile allies known as RYM II, led by Michael Klonsky, had abandoned Weatherman.

We chose October 8 because that was the anniversary of Che Guevara's murder in Bolivia, two years before. Cuba had proclaimed October 8 "El Día del Guerrillero Heróico," the Day of the Heroic Guerrilla. Che was my personal revolutionary saint, as he was for every Weatherman; we wanted his pain and glory for our own. "There's a war going on in the world . . ." began almost every Weatherman flyer, whether printed by the National Office in Chicago or by the local "summer projects" scattered in white working-class neighborhoods in seventeen midwestern and eastern cities. A typical leaflet continued:

All around the world the barricades are going up.
And you're either on one side or the other.
The people who want freedom are all on one side.
The pigs of the world are on the other.
BRING THE WAR HOME!

Right before the July 4 weekend, I drove to Cleveland with Kathy Boudin, an older SDSer who had worked in community organizing and was now working in the Chicago National Office. Our purpose was to convince the National Mobilization Committee, or "Mobe," the huge umbrella coalition that had organized the previous three years' mass protests—in New York, D.C., Chicago, and San Francisco—involving hundreds of thousands, perhaps millions, of people, to support SDS's Chicago National Action.

The Mobe meeting consisted of several hundred peace activists from around the country who came together to decide what demonstrations to plan for the fall. I took the floor in a belligerent mood. I had no time for the Mobe liberals who wanted to hold back the struggle. "The street battles at the August 1968 Democratic Convention in Chicago did more damage to imperialism than all the mass peaceful demonstrations put together," I began. "Now we have to build an anti-imperialist movement in the United States which will be, in effect, the American arm of the National Liberation Front of South Vietnam.

"It will be led by the Black Panthers, the Young Lords, and SDS. The Chicago National Action will help build this revolutionary alliance because it will prove that SDS is a white radical group that will fight."

I ended by demanding that the broad-based coalitions that had until then led the antiwar movement must be willing to follow this new, anti-imperialist leadership. "It is not enough to demand immediate withdrawal from Vietnam; the movement must now *bring the war home.*"

Speaking for the new SDS leadership—but not really for the whole organization—I had purposely excluded the millions of moderate, nonviolent, middle-of-the-road people who now were willing to publicly demonstrate their opposition to the war. I denounced anybody at the meeting who proposed a broader action, with less militancy and violence, as a "liberal," a very dirty word at the time. After all, the liberals Kennedy and Johnson had given us the war in Vietnam. Even those who originally had been inclined to join with SDS in Chicago, like Dave Dellinger, now saw that there was no way the Mobe could participate. My speech was greeted with silence.

A few days after the Mobe conference, I met in New York with Gerry Lefcourt, my attorney, and Martin Kenner, an older grad student who had been arrested at Columbia, friends whose opinions I trusted. They told me they thought the National Action as we had announced it was self-isolating and that we should be mobilizing the largest number of people possible. In other words, the Mobe line again. Martin told me that a diplomat from the Cuban Mission to the United Nations whose job it was to keep track of the North American Left had told him that even the Cubans agreed that the National Action was terrible. Following the lead of the Vietnamese, the Cubans understood the need for the broadest possible unity of as many Americans as possible against the war, not a fantasy of violent revolution in the streets.

Could I discount the Cubans' judgments as reformist and liberal? More, the criticism struck a chord submerged in my own thinking: I was less certain of the efficacy of violent demonstrations than I had realized. I was also secretly relieved to be handed an out: Going up against the Chicago police was suicidal.

When I reported the conversations and my own doubts to Terry

Robbins and Billy Ayers back at the NO in Chicago, they jumped on me. "How could you succumb to that liberal bullshit?" Terry snarled. "We've got to do it; it's the only strategy that will build the revolution." Terry and Billy had been organizing partners in Ohio and Michigan. Terry was the younger—only twenty years old—and smaller, also the more strident and aggressive. He was the hard cop to Billy's soft, sympathetic manner. The scene plays in my memory like a grade-B gangster flick. Billy looks on in smirking contempt as Terry dismisses me with a flick of his ever-present cigarette: "How could you be so weak?" That settled it.

Not only was I easily silenced by Terry's "argument," but I even turned around and used the same ridicule and bullying on others both inside and outside Weatherman. In September, at a public meeting of Columbia SDS, the first of the new school year, I paced back and forth brandishing a chair leg I had just used in a brawl with PL supporters who'd tried to crash the meeting. "You can't be soft and wimpy anymore! You've got to be prepared for the revolution!" I screamed at the crowd of several hundred students in an auditorium in the basement of Butler Library. "I've got myself a gun—has everyone here got a gun? Anyone? No? Well-l-l, you'd better fuckin' get your shit together!"

In the audience I spotted Robert Friedman, my old friend who had been editor of the *Daily Spectator.* I singled him out. "Friedman, are you coming to Chicago next month? You can't sit on the fence, you can't be a wimp all your life!" Robert later told me that was when he realized there was a part of the movement he didn't belong to anymore.

That summer we actually set up tests of our ability to fight—in the guise of "organizing" the potentially revolutionary youth. In Detroit, members of the Weatherman summer project, mostly students from Ann Arbor, ran down a beach brandishing a Vietcong flag with predictable results—a gang of teenagers fought them. In Brooklyn, our people invaded a high school and tied up two teachers while haranguing the terrified ninth-graders about the need to come to Chicago on

October 8 and join the people of the world in their fight against U.S. imperialism. In Cleveland, we started fights at a shopping mall, in which several of our youngest members were arrested and eventually did time in state prisons. To participate in these actions was to prove yourself a hardened "revolutionary cadre."

In early September I drove up to Milwaukee from Chicago to join the local collective in a Sunday-afternoon organizing march. The Milwaukee collective, led by John Fuerst—who had originally founded the SDS chapter at Columbia back in 1966—was made up mostly of students and ex-students from the elite University of Wisconsin in Madison. John had been a graduate student there in sociology after he graduated from Columbia.

About a dozen of us met in a second-floor apartment that served as the collective's headquarters and general crash pad. The local leadership proposed that we march down the main street of a white working-class neighborhood where the "greasers" cruised in their cars. We would win these kids to the revolution by showing them how tough we were.

We had gotten about a block down the crowded street, chanting, "Ho, Ho, Ho Chi Minh, NLF is gonna win!" while waving the blue, red, and yellow flag of the National Liberation Front of South Vietnam and giving out leaflets about the upcoming National Action to startled onlookers, when a group of about a dozen young guys attacked us, going for the flag and, it seemed, me. I took some blows to the body and face, then some kicks to my kidneys. Before I had a chance to react, I blacked out and fell to the pavement. The brawl must have continued around me for a time, but all I remember was a policeman shaking me to consciousness and asking if I needed help. I shook my head, and John and another Weatherman dragged me to a nearby car and drove me back to the collective's apartment.

There we did our inevitable self-criticism, the gist of which was that we should have fought harder to convince the greasers of our seriousness and strength. I don't remember anyone, including myself, voicing the opinion that the whole tactic—and the strategy behind it—was absurd. I didn't even allow myself such a thought.

After the meeting I soaked myself in a hot bath, then managed to drive back to Chicago, where I collapsed. In the morning, bruised and in severe pain, I called Quentin Young, the doctor who headed the Chicago Medical Committee for Human Rights. He told me to meet him at the hospital. I wound up spending three days there for X-rays and observation for possible kidney damage.

I spent a lot of those three days brooding. I knew I was no fighter and suspected that most of the other Weathermen were equally incapable. Still, I was too deeply committed to our course to come up with any new conclusions.

I must have convinced myself that if you talk about something enough, it will come true. When I got out of the hospital, I hit the college trail to try to convince SDSers and other students to join us in October. Besides Columbia, I addressed—or rather attacked—audiences at City College in New York, Yale, Cornell, Purdue, and Indiana University, among others.

One typical speech is preserved in the FBI reports I obtained through the Freedom of Information Act. It seems the Bureau had at least three agents or informers monitoring my appearance at New Haven College in Connecticut on September 26, 1969. I was supposed to have debated a representative of the right-wing Young Americans for Freedom, but when I took the stage, I grabbed the microphone and declared I had no intention of debating with a lackey of the imperialists. According to one FBI informant, I berated the United States for its involvement in Vietnam and its oppression of minority groups, including especially the political repression of the Panthers and the jailing of their leader, Huey P. Newton.

The core of my talk, however, was a description of the impending National Action. "SDS plans to have fifteen thousand students demonstrate in Chicago during the trial," I reportedly said. "In Chicago the pigs have to be wiped out. We're going to fight with violence and wipe out Chicago. Demonstrations are over, violence is the only way; there just ain't no other way. A lot of people are going to have their minds blown."

One agent reported that my speech was difficult to follow because there was little or no continuity and there was considerable heckling and jeering from the audience. He also said that my speech was filled with obscenities and vulgarities. Apparently I had six or seven bodyguards around me, standing with their arms folded, and when I finished (hurriedly because of the heckling), we marched out of the theater shouting, "Ho, Ho, Ho Chi Minh, NLF is gonna win!"

At a talk three days later at Indiana University, the FBI records, all too accurately, that I declaimed:

> Our final goal is revolution—armed struggle.
>
> Come to Chicago and fight!
>
> Next week, October 8–11, some people will get hurt, some killed, to build the revolution.
>
> We want whites to take risks now—affinity groups will be the main tactics. Whites in twos and three will off the pigs.
>
> Now we got to kick *[obscene]* in this country—got to create disruption
>
> Don't have non-violent marches.
>
> Chicago will be hit by a shock wave.
>
> We will close schools, cause strikes, etc.

Toward the end of my harangue, I'm alleged to have said, "I hate the SDS, I want it to die. The SDS is not serious enough." Presumably Weatherman was.

Just as the absurdly macho Weather actions such as fighting white working-class kids on streets and beaches were repeated in Chicago, Brooklyn, Cleveland, Columbus, or Seattle, the internal culture of Weatherman reproduced itself flawlessly. Collectives were always located in cheap apartments in deteriorating, mostly-white neighborhoods and inhabited by a dozen or more young men and women, some runaways or street kids, but primarily ex–college students. There were always mattresses and sleeping bags on the floor, dirty dishes in the sink, clothes strewn around haphazardly. In the drive to become revolutionary fighters, we tried to jettison our middle-class values, including all personal possessions. Privacy and cleanliness were shed as well: If you had only one set of clothes, how could you wash them? And if you had more, your clean clothes would probably be appropriated by a brother or sister who needed them. Once I opened a closet door at the Brooklyn collective house and found a remarkable sight: It was filled with what seemed like dozens of women's discarded portable hair dryers, relics from a former life.

The collectives determined each person's daily schedule, which consisted mostly of meetings and organizing in the neighborhood—perhaps going out to high schools to distribute leaflets about the National Action or hanging on street corners rapping with kids. Very few people had the time or inclination to work at jobs. The collectives supported themselves, in a manner of speaking, through individual contributions from the members' savings or quite often from their indulgent or lied-to parents (staged weddings were common). A handful of collectives with more working-class members and therefore fewer family resources developed elaborate burglary and fencing operations, however. To the leadership this was a sign of the "proletarianization" of our organization.

The heart of collective life was the "criticism, self-criticism session." The group would choose one person, attacking that individual's work and attitudes with the goal of helping him transform himself into a Communist revolutionary. Individualism, ego, and weakness had to be rooted out. We had discovered this method from *Fanshen,* by William Hinton, an American who lived and worked in China. The book was required reading in all collectives. Hinton chronicles the ups and downs of the revolution in a Chinese village right after the 1949 liberation from the Nationalists; he describes in detail how the "criticism, self-criticism, transformation" sequence was used by the Communist Party cadres to remold themselves and become more effective in the struggle for transformation of the village.

One of our favorite passages from *Fanshen,* which we often quoted, justified our attempts to remake ourselves:

> The human consciousness may be compared to an artichoke. Its tender core is enclosed in layer upon layer of defenses, excuses, rationalizations, and approximations. These must be peeled off if one is to discover the true complex of motives driving any individual. . . . What made self-revelation possible for the work team members . . . was the deep commitment every one of them had to the success of the land reform movement. They freely examined themselves and their comrades, not for partisan advantage, not for the sake of exposure, not as an exercise in mea culpa, but in order to remove obstacles in the way of more effective work.

I remember one day walking into a criticism, self-criticism session at the collective in Park Slope, Brooklyn. A young man, a gentle, thoughtful person whom I had known at Columbia SDS, was on the hot seat. He had violated some collective rule, such as needing permission from the collective to go and see an old friend or family member. The poor guy in the end broke down under the onslaught: Yes, he admitted, he was a bourgeois; yes, he had doubts as to whether we would win; yes, he was a racist and sexist down deep.

I knew that the whole thing was nuts but couldn't intervene to stop it. I had mixed feelings, recognizing the inherent unfairness of a group attacking an individual. How much better was each of us than the poor guy being criticized? Wasn't his doubt about the possibility of achieving a thoroughgoing personality regeneration in a matter of months quite reasonable? On the other hand, I believed, as much as anyone else, perhaps more so, in the need to harden ourselves through group criticism. So I said nothing.

I did not realize at the time that we had unwittingly reproduced conditions that all hermetically sealed cults use: isolation, sleep deprivation, demanding arbitrary acts of loyalty to the group, even sexual initiation as bonding. It's strange that these practices can arise without any conspiratorial mastermind or leadership cabal.

Within two days the young man had left the Brooklyn collective in despair over his weakness. The remaining cadre just chalked it up to a necessary hardening and purifying process—only the best (ourselves) would survive.

Very quickly that summer, a new and startling "line" emerged from the Weatherwomen: "Monogamies" had to be smashed! Exclusive relationships between men and women were now seen as reinforcing the old patterns of female subservience; women had to break out of the old ways, stand on their own two feet. Simultaneously, in several collectives, women observed that couples tended to protect each other in criticism, self-criticism sessions. So "the deal" of mutual self-protection had to go.

In theory, women liberated from these monogamous relationships would be free to grow, to get stronger, to struggle, to be "up front" at demonstrations, to become self-reliant. As one Weatherwoman put it, "Our monogamous relationships broke up because we simply didn't need them anymore." The new line was adopted on the last day of a Weatherman conference held in Cleveland over Labor Day weekend in 1969. Women from the Detroit collective explained the question for the whole Weather organization, several hundred young people from around the country gathered at a church in one of the black ghettos. They laid out their main argument, that abolishing monogamy would liberate women from the old definitions based on attachment to men.

The only opposition came from Michael Klonsky, now leader of the rival RYM II faction, who was present at the conference. He was soundly booed when he asserted that the family was the basic unit of revolution in every socialist country. Of course, Klonsky was right. In no revolutionary society has the family been abolished; that, historically, has been the false charge of all reactionaries, that Communists want to "hold women in common."

On the other hand, "smashing monogamy" was a response to a real problem—how are women going to make great gains in their liberation very quickly and become full revolutionaries, not subordinate to men or a man? The solution for many New Left women was to separate from the men and join all-female organizations. For the Weatherwomen the solution was to abolish monogamous relations yet stay within the integrated anti-imperialist revolutionary organization.

This was one of the few cases where a new line was not initiated by the Weather Bureau, which was how the leadership collective instantly had become known. The reason was simple: Except for Bernardine Dohrn, the Weather Bureau was all men. Anyone who'd ever met her would recognize Bernardine as a star—intelligent, beautiful, commanding in a unique way through her warmth, empathy, and passion. I was proud just to be in her company. No other woman even approached her in her charisma.

After I was elected SDS national secretary in June, I had moved to Chicago from New York with my girlfriend, Jennie. She had gone into the Chicago collective, while I was based out of the National Office. After the Cleveland conference, we returned to Chicago and spent a night crying in each other's arms, seeing no other choice than to go along with the new line. We agreed with the need to become revolutionaries, yet we didn't want to give up each other. The next morning Jennie moved out of our apartment into one of the collective houses. We were now free to be proper revolutionaries.

For me it meant freedom to approach any woman in any collective. And I was rarely turned down, such was the aura and power of my leadership position. For the second time in my life, my fantasies were being fulfilled: I could have almost any of these beautiful, strong revolutionary women I desired. It was a moment of extreme sexual experimentation. Group sex, homosexuality, casual sexual hook-ups were all tried as we attempted to break out of the repression of the past into the revolutionary future.

On one ride from Chicago to Detroit, all fourteen or so of us, except perhaps the driver, writhed naked on the floor of the van while hurtling down the interstate, legs, arms, torsos, genitals interlocked with no particular identity attached. A strange, truly disconnected experience, not one I care to repeat.

In Cleveland, during the collective's morning karate workout, I noticed a beautiful young blond woman, a recruit straight out of high school. I'd pretend an attack, and she would defend and counterpunch with glee. We looked at each other in that undeniably flirtatious way. That night we had sex together in a closet on a bed of dirty laundry. I was shocked when I noticed that she bled all over the clothes—she was a virgin. I've often wondered what that experience meant to her, since we never talked.

The line on homosexuality was that women should get close to women, men to men. In a revolutionary collective, there should be no barriers between people. Mainly, women slept with women; this was openly allowed. Men were more repressed. My own internal taboo never let me do more than fondle JJ's penis during a threesome with a Weatherwoman one night. This homosexual experimenting had nothing to do with gay liberation. The organization was essentially male-oriented and macho, and a large number of gay men and women never came out of the closet in Weatherman. I only learned about their suffering years later.

In public, and to people outside the organization, I explained smashing monogamy:

"Since sex is the ultimate intimacy in human relations, Weathermen are building political collectives bonded with this intimacy among all our members, not between monogamous couples," I would argue. I forgot that intimacy is of necessity a result of uniqueness and exclusivity—which exist outside the problem of possession.

Our sexual ideology quickly proved to be disastrous. Smashing monogamy drove many good people out of the collectives. Special relationships were criticized by the group as being corrupt. Instead of bonding people into a true collective, what we were doing created fragmentation and alienation. As I made my rounds of the collectives, for example, women turned against each other because they were competing for my attention. They wanted to be close to power. The situation was inherently unstable.

Weather sex was also a disaster for medical reasons: Gonorrhea, pelvic inflammatory disease, crab lice, and a nonspecific genital infection we called "Weather crud" were epidemic among us. Once, while making love with a woman, actually an old friend from Columbia, I looked into her face and saw a crab in her eyebrow. I could not go on. Sympathetic doctors treated us with antibiotics and other drugs, yet we continued throughout the latter part of 1969. By early 1970, the spasm of indiscriminate sex was over. We were exhausted, and few of us, to my knowledge, ever looked back on the sexual experimentation with nostalgia.

However, women did come into the leadership of each local Weather collective. This was something new in SDS and in the movement generally. But just as women in business have to be tougher than men to achieve equal success, the Weatherwomen outdid themselves to prove they were revolutionary fighters.

Two days after the Cleveland conference, about seventy-five Weatherwomen descended on a high school in Pittsburgh for the "first national Weatherwomen's action." They ran through the halls screaming "Jailbreak!" They broke in to classrooms to exhort the kids to "join the revolution and come to Chicago next month," then ran headlong into squads of Pittsburgh police who had been called to the school. Twenty-six women were arrested for riot and incitement to riot. I'd be amazed if even one kid from that high school came to Chicago. The unacknowledged purpose of the action was internal: The women had now proven to themselves and to the men in the organization that they were committed revolutionaries.

VIETNAMESE WOMEN CARRY GUNS! read an inscrutable wall slogan painted at night in every city with a Weather collective. As little as this must have meant to the average person passing by the next morning, to the Weatherwomen who organized the clandestine painting action it marked an enormous passage. How much further could a woman get from the roles assigned by our culture—wife, mother, boss's girl Friday—than identification with the image of a tiny Vietnamese woman, AK-47 in hand, defending her country and family from the huge male American invaders?

The summer of 1969 had brought the unusual opportunity to meet with "the enemy." SDS, and specifically the Weathermen, were invited to join a large delegation from the antiwar movement who would meet in Havana with representatives of the National Liberation Front of South Vietnam and of the Democratic Republic of Vietnam (North Vietnam). We sent fourteen Weatherpeople, making up almost a third of the whole delegation. Bernardine Dohrn led our contingent; I stayed behind in Chicago to continue organizing for the National Action.

The purpose of the conference was to explain to members of the U.S. peace movement the significance of the founding of the Provisional Revolutionary Government of South Vietnam the month before. At the high point of U.S. troop occupation of Vietnam, a broad government was in place, ready to assume power once the Americans withdrew. According to the Vietnamese, the U.S. military was in the process of being defeated: "The U.S. can never escape from the sea of fire of people's war." But to convince the Americans to withdraw, the antiwar movement in this country had to become even broader and more committed. That should have been our job.

Our delegation had to return through St. John, New Brunswick, Canada, because it wasn't possible to travel from the United States directly to or from Cuba. JJ and I went to meet Bernardine and the others who had traveled home on a Cuban freighter. At an all-night session in a small hotel room, we browbeat Bernardine into conforming to the Bring the War Home line that had been developing all summer. Excited, we told her about the collectives, the Weather actions, our growing cadre development, the criticism, self-criticism. She was torn: In Cuba the Vietnamese had talked with the group about not getting too far ahead of the masses, about revolution developing as a growing consciousness in people's minds, step by step. They had also made it quite clear, she told us, that above all, the Vietnamese needed unity of the American antiwar forces to end the war.

But we were Bernardine's closest comrades, the ones she had united with to split SDS. Our idea of the revolutionary youth joining with the people of the Third World had yet to be tried. We'd been on a roll, and now, as a result of this one trip, she couldn't turn the whole thing around.

It wasn't a fair fight: JJ could outtalk anyone. Bernardine capitulated and told us what a Vietnamese comrade, perhaps idly, had told her: "When you go into a city, look for the person who fights hardest against the cops. That's the one you want to talk all night with." Over the next months, this became part of the Weather canon, repeated endlessly. Despite all our talk about "following Third World leadership," we were following our own crazy instincts. "It's time to kick ass," we told ourselves.

On a hot Saturday in early August, at the Mall in Central Park, I was among a contingent of several dozen New York Weathermen who marched up onto the stage at a rally commemorating the twenty-fourth anniversary of the bombing of Hiroshima and Nagasaki. We elbowed aside the surprised organizers from the Fifth Avenue Peace Parade Committee and seized the microphone. Jeff Jones, my comrade in the Weather Bureau, commenced to rant at the several thousand peace protesters assembled. "It's not enough to be for peace, like the liberals who organized the rally; you've got to be for revolution. And that revolution will be violent, so get ready." As he continued, the audience just drifted away; we had destroyed the rally more effectively than any right-wing counterdemonstrators or government agents might have.

Seventeen years later, while laying bricks at a farm cooperative in revolutionary Nicaragua, my work partner, a mason from Santa Fe, New Mexico, described to me the moment when he had left the antiwar movement, starting a long period of political inactivity that didn't end until 1986, when he joined the New Mexico Construction Brigade to Nicaragua.

"It was at a Hiroshima Day rally in Central Park in New York, in the summer of 1969," he said. "A bunch of crazies took over the speakers' platform and began screaming about violent revolution. It seemed like most of their hate was aimed at the other people in the movement. I got scared and left; I guess I didn't want any part of a movement which had so much hatred and violence."

After a deep breath, I admitted I was one of the crazies responsible for the takeover of that rally. "It's among the biggest regrets of my life," I told him.

Two days before the National Action was to begin, I found myself in New York City. The next day I would use my forged half-fare youth card (I had turned twenty-two, past the cutoff age) to fly back to Chicago. That night I was to meet Gerry Long, my Weather Bureau comrade, about five years older than I and a hard-core anti-imperialist, in Greenwich Village for supper. Both of us were as nervous as soldiers awaiting the next day's battle. Gerry must have seen something in my face. He said, "Yeah, I'm scared, too."

True to form, Gerry and I decided to splurge on our Last Supper. I had about fifty dollars of the organization's money in my pocket, so we went off to a fancy Italian restaurant in the Village. We talked of the organization and the action coming up. Gerry was a realist, and that night he expressed his doubts. "Is this worth dying for? Is this a real battle of the revolution?" All I could give him were the truisms we were locked into, so I didn't. I was in a deflated mood that night, bordering on depression. We decided to get drunk on red wine; what else can you do when there are no answers?

Gerry told me a story: "On the trip to Cuba this summer, two of us were able to slip away with one of the Cuban guides. We were told that some comrades from the Foreign Ministry wanted to talk with us away from the rest of the group, which probably contained FBI informers. We thought they wanted to talk about Weatherman.

"We were taken to a small, middle-class hotel in the same Havana barrio as our hotel. It turned out to be Fidel's home. We had all been dying to meet Fidel. You know, he really is big; I guess that's one reason they call him 'El Caballo' [the horse]. He asked about the conference, and then he talked for a while about the American military in Vietnam—how they would inevitably lose. It was strange: He wasn't really paying us much attention. His mind definitely seemed to be elsewhere.

"Finally he just stopped, and we were quiet for a short time. Then he blurted out, 'You know, something very troubling has just happened. A friend in the Bolivian government sent us a package; it arrived yesterday. Che's hands, his very hands, preserved. Before they destroyed his body, they chopped off his hands to prove they had him. They were definitely his hands, I recognized them. I don't know what to do with them.'

"Fidel was just at the point of tears. He shook himself out of it, stood up, thanked us for coming. The interview was over."

Both of us were quiet for some time. Finally, after all the public heroics and underneath the glory of revolutionary war, is human, individual death. The meeting with Fidel was not at all what Gerry had anticipated.

The next day we were in Chicago.

10

Days of Rage

When I arrived in Chicago, the first thing I learned was that the police had just murdered a sixteen-year-old Puerto Rican youth. Before the end of the weekend, they would kill his nineteen-year-old brother, a GI home on leave to attend the funeral. Over the summer both the Black Panther Party's office and the SDS National Office had been raided by the Chicago police, with several Panthers arrested. A half dozen Panthers had died in jail and on the streets.

Two weeks before, twelve members of the Chicago Weatherman collective and people from the National Office had been beaten and arrested as they attempted to hold a rally and a press conference outside the courthouse where the Chicago 8 trial was beginning. Warnings of the coming radical mob violence were now appearing regularly in the press, along with declarations from the Chicago Police Department that they were ready for us.

Not shrinking from the fight, the Weather Bureau sent a clandestine squad to blow the ten-foot-high bronze statue of a policeman off its pedestal in Haymarket Square, the site of the historic Haymarket Massacre of 1886. The logic of this first Weather bombing seemed crystal clear to us—blowing up a statue linked us to the nineteenth-century revolutionaries who had fought the police.

The president of the Chicago Police Sergeants' Association was quoted in the paper. "We now feel," he said, "that it's kill or be killed."

A poster appeared on Chicago walls that week, a product of the National Office print shop. It read, modestly, "During the 1960's, the American government was on trial for crimes against the people of the world. We now find the government guilty and sentence it to death in the streets."

The Weather Bureau, our generals, met to map the battle plans. We divided up assignments: I wasn't to go into the streets until Saturday afternoon at our big, final, downtown Loop action. Before then I was to keep myself unbusted and to help organize support, especially raising bail money.

The early evening of October 8, a few hours before the first demonstration was to begin, Terry Robbins dropped me off at a large, upper-middle-class house in Wilmette, a rich North Shore suburb. It was owned by friends of his I didn't know who were away. I sat in the big, fancy kitchen, alone with my anxiety and my elation, listening to the FM radio for news of the action.

The news that came in was that only a few hundred people had gathered that first night in Lincoln Park, the site of the big police riots against the demonstrators at the Democratic Convention the year before. It was a much smaller crowd than we had anticipated—only two hundred Weathermen dressed in combat boots, leather or denim jackets, some with gas masks and almost all armed with sticks or pipes or rolls of pennies in their fists. Chilled by fear and the October night, people broke up park benches—it was war, after all—and lit a bonfire. They tried chanting, "The revolution has come! / Off the pig! /

Time to pick up the gun! / Off the pig!" but it didn't seem to boost morale. Some incendiary speeches were attempted, but dejection over the small numbers plus fear about the violence to come put a huge damper on the rally.

Bernardine Dohrn spoke at about 9:00 P.M. "This is the second anniversary of the death of Che Guevara," she reminded people. A large banner in the crowd showed his famous portrait, under which we had written the single word AVENGE! Fallen martyrs are always good for motivation. An hour later Tom Hayden appeared. A founder of SDS who'd drafted the Port Huron Statement, a leading organizer of the Democratic Convention demonstrations the year before, he was now one of the Chicago 8 on trial. It was cheering that a non-Weatherman supported our efforts. Abbie Hoffman, another of the defendants, had also come along.

Tom told the crowd, "It's a lie that we oppose this Weatherman demonstration. It's good to see people coming back to Chicago, back to Lincoln Park. We welcome any intensification of the struggle." Then he and Abbie departed.

Very few other organizations had supported the National Action. Only one other national group, Youth Against War and Fascism, the youth branch of the Workers World Party, one of the many tiny Communist parties, sent a contingent, while from Chicago only Rising Up Angry, the organizing project in Uptown, a mostly white, Appalachian neighborhood, sent a token delegation. The next day Fred Hampton, the chairman of the Chicago Panthers, would make a speech at a rival demonstration called by Klonsky's RYM II.

In his characteristically eloquent style, Chairman Fred was clear. "We support RYM II only. We oppose the anarchistic, adventuristic, chauvinistic, individualist, masochistic, and Custeristic Weathermen. We don't dig confrontations that lead people into struggles they're not ready for." Among ourselves the Weather Bureau dismissed his attack, saying that he didn't understand what we needed to do to organize white people.

Just as dispiriting, only a tiny handful of the more than 350 campus

SDS chapters were represented. Weatherman was alone. Hayden's benediction probably afforded little warmth to the few hundred there. I was experiencing that loneliness out in Wilmette, though I was a lot safer and warmer.

After Hayden, Jeff Jones took the megaphone. He stood in full combat gear—leather jacket, gloves, work boots—as he introduced himself, "I'm Marion Delgado." The cadre cheered and gave that high-pitched keen, or ululation, that came from the back of the throat, which we had learned from the film *The Battle of Algiers.* Marion Delgado was the name for the archetypal Weatherman hero. Earlier in the summer, the editor of *New Left Notes* had run a picture of the real Marion, a grinning, malevolent Dennis the Menace type kid, holding up a piece of concrete almost as big as himself, to show how he had derailed a passenger train. This was an actual incident in California in 1947. We had adopted the satiric slogan "Marion Delgado, live like him!" As long as we were partisans of the politics of transgression, it was fitting that we chose a demonic five-year-old vandal as a role model.

That night, hearing the name was the prearranged signal for the riot to begin. "Marion Delgado" announced that Judge Hoffman of the conspiracy trial lived at the Drake Hotel on the fashionable Gold Coast, "a big fancy pigpen down on Michigan Avenue, and Marion Delgado don't like him, and the Weathermen don't like him . . . so . . . let's get him!"

The park was loaded with plainclothes cops lurking on the fringes of our rally; outside the park, hundreds of uniformed police troops waited in reserve. But the crowd streamed out of the park into the streets before the cops could organize themselves. People were trotting in a tight group, chanting, "Ho, Ho, Ho Chi Minh, dare to struggle, dare to win!" Then the first window was broken. It was now a mob racing through the streets of the exclusive Gold Coast neighborhood, attacking a lone Rolls-Royce, hitting plate-glass windows in banks, hotels, and restaurants. After about four blocks, the police managed to set up a barricade. The mob turned down a side street, only to find themselves running into a line of helmeted police.

Some of the leaders in the front ranks picked up speed and sprinted through the line, but police clubbed many more Weatherpeople to the ground. Others ran off in a different direction, still trying to get to the Drake Hotel and Judge Hoffman, whom we had declared the personification of the enemy. The police went on the attack now, beating any individuals or groups they found, going for any unprotected body parts. Kirkpatrick Sale, in his history *SDS,* quotes one Weatherman, "Bodies were just mangled, people were bleeding out of their mouths and their noses and their heads. . . . It was vengeance." Police fired shotguns, pistols, tear gas at the demonstrators; some even drove their squad cars full speed into running crowds.

After just an hour, the demonstration—and the carnage—was over. The result: six Weathermen shot, many dozens more injured, sixty-eight arrested; twenty-six policemen were injured, though none seriously. Our people straggled back to the "Movement centers," churches loaned to us by sympathetic clergy. They were scared and proud at the same time, and still not defeated.

For the Weather Bureau, it was about what we had expected. Early the next morning, we planned the subsequent actions. We canceled some parts of the three-day schedule— the rock concert (billed as the "Wargasm, a revolutionary youth culture celebration"), a march on the federal courthouse where the Chicago 8 conspiracy trial was under way, and several high-school "jailbreaks"—because we'd realistically reassessed how few and weak we were. But for the most part, we just put our heads down and kept charging forward.

We decided to continue that morning with the Women's Militia action at the military induction center in the Loop. It was considered one of the most important events of the three days: Women fighting for revolution would set an example for all women. About sixty or seventy women showed up in Grant Park downtown, chosen as the rallying point because of the large battle that had taken place there in 1968 at the Democratic Convention. They were prepared with their helmets, gloves, and clubs, singing, "We love our uncle Ho Chi Minh, deep down in our hearts, / We love our Chairman Mao Tse-tung, deep down in our hearts," and chanting, "Oink, oink, bang, bang! Dead pig!"

Bernardine spoke from the center of the crowd. "For the first time in history, women are getting themselves together. We're not picketing in front of bra factories. A few buckshot wounds mean we're doing the right thing. This is not a self-indulgent bullshit women's movement. We refuse to be good Germans. We live behind enemy lines."

Then the charge—trying to break out of the park, into the street to get to the induction center. It was a suicide mission—the women ran right into a double row of cops; several, including Bernardine and Kathy Boudin in the front ranks, were wrestled to the ground, beaten, then thrown into vans while they chanted, "Power to the people!" The rest were ordered to drop their clubs, helmets, gloves, while the police escorted them to a nearby subway entrance in defeat.

At a large assembly that night, we formally redefined the National Action as a "cadre action," instead of the "mass action," involving thousands, that we had originally expected. The few and the brave. But we were vulnerable, easily identified and picked off. That day the police had been grabbing Weathermen off the street, charging them with riot, assault, and other felonies from the night before. Also, the Illinois governor, Richard Ogilvie, had called out twenty-five hundred National Guard troops; it was possible there would be more shooting.

In a show of bravado at that Friday-night meeting in the basement of a church, I argued that we should go on with the big demonstration set for the next day. This would prove we weren't defeated, that we were serious about revolution, that we lived by Che's law, "In revolution you either win or die."

That night there was considerable criticism of the Weather Bureau for bad leadership—exaggerating the numbers that were coming, not being good street fighters; also, several individuals admitted at the meeting that they were scared and were dropping out. Despite the naysayers, in the end it was decided that we would go ahead with the Saturday action.

During this time I was operating mechanically, efficiently, focusing on whatever task was in front of me in order to mask my inner confusion. In reality I was in a state of shock over the one-sided police violence and our obvious failure, but I just plunged ahead, not admitting anything. We could not appear defeated.

I had spent all Friday afternoon at a booth in a Chinese restaurant in Evanston, making calls around the country trying to raise bail for the dozens of our people in jail. A Weather leader from Michigan had run in looking for me. He had recognized an undercover cop from the Chicago police at a meeting in the church basement, a guy who'd once arrested him. Weatherman being a top-down organization, he felt obliged to ask a Weather Bureau member what to do about the cop.

I said, "Get rid of him," meaning throw him out.

About ten minutes later, several Weathermen came running into the restaurant. The cadre leader had beaten up the undercover cop really badly. The minister of the church had come in on the scene and ejected us from his basement. I felt terrible that I hadn't made myself clear. The situation created another crisis: we now had to help the first Weather fugitive go into hiding. Of course, he took on the code name Marion Delgado.

Very early Saturday morning, a hundred police raided the remaining church basements in Evanston, and forty-three of our people were dragged off and arrested. This generated total confusion, as people scurried around trying to find shelter from the police and the cold Chicago rain. I got no sleep while I worked to try to pick up the pieces of the "revolutionary army." I was supposed to lead the troops into battle at 2:00 P.M. that day.

The old Haymarket Square was a large, open area of converging streets and abandoned trolley tracks, surrounded by dilapidated industrial and commercial buildings. In May 1886, it became the historic site of the Haymarket Massacre, where eight police were killed by a bomb blast when they tried to break up a union organizing meeting. There followed a trial of eleven union leaders who were framed and convicted of murder, eventually hanged. The international workers' holiday, May Day, originated in commemoration of the event; it continues to be honored worldwide, except in the United States. We chose Haymarket as our rallying point in a self-conscious attempt to link the present revolution with American anticapitalist history.

At 1:45 P.M., fifteen minutes early, I and about six other cadres approached the empty pedestal—its statue of a policeman having been knocked off by our bomb just days before—bearing the inscription IN THE NAME OF THE STATE OF ILLINOIS, I COMMAND PEACE. I was cleverly disguised with a dime-store mustache. As other Weathermen began to assemble, a squad of cops, equally cleverly disguised as longshoremen heavies, rushed us with clubs and mace. I grappled with one of them, whom I recognized as Maurey Daley (rumored to be old Mayor Richard Daley's nephew) of the Chicago Police Department's Red Squad, our nemesis. When the mace hit me, I decided to stop fighting and just curl up into a ball on the ground to protect myself, especially my eyes. Daley did a double take, surprised that I hadn't fought back. The whole battle took about a minute, from clubs to handcuffs.

My comrades and I were thrown into unmarked cars and taken to a station for booking. I was amazed to find I wasn't hurt too badly. The mace hadn't affected my skin at all, though it did burn my eyes.

At the station I was charged with riot, assaulting an officer, and resisting arrest. I was somewhat disappointed that I hadn't seen any street action, but I was also relieved that my part in the battle was over so quickly. In the next hours, my comrades were placed under arrest and brought in. I learned that JJ had addressed the rally in my place, saying, "People know that we are marching about what white people have to do. . . . We don't really have to win here. . . . Just the fact that we are willing to fight the police is a political victory."

The march had left Haymarket on the approved route, then had broken out of police lines and rampaged through the crowds of Saturday shoppers. Again the wild slashing at store windows and cars with bricks, pipes, clubs, chains. Again the counterattack by the police, also with clubs and blackjacks. Scores of our people were injured, including one Weatherwoman who had both arms broken and another who had most of her teeth knocked out and her jaw broken. The two of them were both friends of mine. Demonstrators were piled in heaps on downtown streets as they were subdued by the cops.

A total of 123 people were arrested that afternoon, perhaps half of all the marchers. On the other side, thirty-six cops were injured, though only one seriously: the city's assistant corporation counsel, an attorney named Richard Elrod, who habitually joined the police in anti-demonstrator rampages, had attempted to tackle a Weatherman and had smashed into a wall instead. Brian Flanagan, from Columbia SDS, was charged with attempted murder but was later acquitted. Elrod has remained in a wheelchair for the rest of his life. Years later he was elected sheriff of Cook County.

Cook County Jail was overflowing with the addition of almost three hundred Weathermen, the total number arrested over the three days. Inside, we nursed our wounded and held discussions, as usual, about the demonstration—its value, the tactics, lessons learned. Though we were exhausted, we were also elated that we had actually done as much damage as we did. That was our goal—a white riot—and we had produced the headlines we wanted: RADICALS GO ON RAMPAGE and GUARD CALLED IN CHICAGO AS SDS ROAMS STREETS. When the press began referring to the National Action as the "Days of Rage," we proudly accepted the new name.

The next morning I was one of the first bailed out, befitting my exalted status on the Weather Bureau. We eventually raised $2.3 million in bail bonds, which is actually $230,000 in cash, largely from relatives and friends. Most everyone was out within a month, though one or two, "the least important," did stay up to four months in Cook County Jail waiting for trial, then were sentenced to time served. Weatherman was nothing if not hierarchical.

Our legal strategy was to tie up the courts with jury trials, forcing the state to prove each and every charge, which of course it couldn't. Many of the arrests were made in sweeps, in which likely-looking suspects were picked up by the cops. My own bust was actually a preemptive strike by the Red Squad, since I was just standing in a public square waiting for a legal rally to start when I was jumped by a gang of thugs who later turned out to be plainclothesmen. My case didn't come to trial, since it was repeatedly postponed, and then by next spring I couldn't be located.

Less than two weeks after the end of the Days of Rage, the Weather Bureau got together at White Pines State Park in northern Illinois. Though it was a clear, brisk fall weekend, most of our time was spent inside the larger of the two wood-frame cabins we rented, reviewing where we'd been and charting the future of our organization. Terry Robbins was chain-smoking, sprawled out in a chair or on a bed, and continually extended his arm and dropped the ashes on the floor. By the third day, the floor was deep with ashes and other debris.

Bernardine opened the meeting with a summary of the National Action. Her view was that the action, despite its costs, had been a success: We had begun the war against the pigs. This first skirmish naturally was our Moncada, which would inspire future battles. Moncada was the name of a Cuban military barracks in Santiago de Cuba, which a small group of revolutionaries, led by Fidel Castro, attacked on July 26, 1953. Despite being a military defeat—most of the rebels were either captured or killed—the Moncada attack began the process that six years later culminated in the overthrow of the dictator Batista. In fact, Castro's organization subsequently took the name "The Twenty-sixth of July Movement."

Moncada stands in revolutionary history as a great example of defeat transformed into victory. It's a perfect symbol: The attempt itself, not its immediate outcome, is what's important. The mere fact that someone was willing to go up against Batista's guns inspired other Cubans to join the revolution. Moncada as historical precedent and legend kept us from despairing; in the trajectory of revolution, defeat always precedes victory. We would build on our defeat in the streets of Chicago, just as the Cubans had, and go on to revolutionary victory.

At one point I timidly mentioned the huge disparity between our predictions of how many would come to Chicago—fifteen thousand at our most optimistic moments—and how few actually showed up. There had been talk of a trainload of revolutionary youth coming from Detroit. Terry took my question as an attack. "Look, the five hundred who came to Chicago are the beginning. Of course it'll start small; how many white radicals in or out of SDS are actually revolutionaries? People are so bought off with white-skin privilege that it's amazing that so many came. But this is a start.

"Why are you always so pessimistic? Don't you think we can win?" he bullied.

Had I opposed the line, I would be repudiating all that I'd fought for over the last year, plus I would be turning my back on my comrades, for whom I felt a deep loyalty. We were the real revolutionaries, the ones who best understood the crimes of imperialism, the ones who were willing to fight. Where else would I go, whom else would I join with? So I backed down, "No, I was just concerned with how we could have been so wrong in our estimates." My question fell like a lead weight. No one picked it up.

Bernardine was speaking now. "This is what we've been waiting for." I honestly didn't know what she was getting at. "The next step after the National Action is to move to a higher level of struggle, to build the underground. Street violence is an unsustainable tactic—it makes us too vulnerable and costs too much. We've got to be able to work clandestinely.

"We've learned from Che that the only way to make revolution is to actually begin armed struggle." Over the previous summer, the Weather Bureau had made sure that everybody in the collectives had read and understood Régis Debray's *Revolution in the Revolution?* Debray had summarized the thinking of Fidel and Che in Cuba about how revolution was going to happen in Latin America. They generalized from the Cuban experience of a small guerrilla band going to the Sierra Maestra and beginning armed operations against the dictator's army, gaining victories, and growing into a full revolutionary army that eventually seized power. In this model the guerrilla nuclei, called *focos* in Spanish, attracted campesinos as fighters and supporters; organizing in the cities among workers and middle-class people was secondary to the actual fighting.

The implication of the *foco* theory was that all previous models for revolution had proved ineffective and were now superseded. We didn't stop to notice that Che had already died in Bolivia using this strategy; his guerrilla column was isolated from the indigenous people of the jungle, his men looked on as foreigners, and was destroyed by the Bolivian army. *Foquismo* wasn't working anywhere else it was then being tried, such as Brazil, Uruguay, Argentina, and even outside Latin America. But we remained fundamentalists to Che's line. When you have the truth, as handed down to us through Debray, it takes a lot to reject it.

Although I believed in the *foco* theory, fear shot through me when I heard Bernardine talk about actually beginning the clandestine armed struggle. Going "underground" would mean not only a profound shift in the organization but also a complete transformation of our lives, yet she had spoken coolly, rationally, as if she were suggesting we go out for supper. She proposed we had to make two fundamental decisions in this meeting: that a "front four"—Bernardine, Jeff, JJ, and Terry—be given the go-ahead to plan the clandestine work, and that the rest of us would, over the next few months, close the National Office, abandon SDS, and take the entire Weather organization underground.

There was amazingly little debate on this momentous decision, as if the whole thing were already decided because it was so obvious. Nobody even said "wait a minute." A completely new line—abandon SDS and build the underground—had just emerged in this meeting, and with it a new leadership. No longer would we function as a coequal collective; we were now led by a clique within a clique.

The future revolutionary cutting his bar mitzvah cake, 1960, age thirteen, in Maplewood, New Jersey. My parents, Bertha and Jake, and my maternal grandmother, Ella Bass, are behind me watching; my paternal grandmother, Mary Rudnitsky, foreground, is smiling for the camera. I'm wondering about the existence of God.

Sue LeGrand, my wife and partner in the underground, in 1967, about the time we met at Columbia University. She was twenty-one, a Barnard transfer student fresh from Missouri. We spent seven and a half years together as fugitives; our first child was born underground.

A noon rally called by SDS and SAS on April 23, 1968, sparked the eventual takeover of five buildings. The action was intended to protest the discipline of the IDA 6, who are standing on the Sundial. By 1:00 P.M., roughly eight hundred people had gathered.

Columbia College Acting Dean Coleman and me in Hamilton Hall, just before he became our temporary hostage. It was the improvised beginning of what became the occupation of five buildings on April 23, 1968.

JOIN US!!

SIT-IN: HAMILTON HALL!

WE'RE STAYING UNTIL THE FOLLOWING DEMANDS ARE MET:

1. All disciplinary probation against the six originally charged must be lifted with no reprisals.
2. Kirk's Edict on Indoor Demonstrations be dropped.
3. All judicial decisions should be made at an open hearing.
4. All relations with IDA must be severed.
5. Construction of Columbia gym must stop.
6. University see that all charges against persons arrested for participating in demonstrations at gym site be dropped.

JOIN JOIN JOIN JOIN JOIN JOIN JOIN JOIN JOIN JOIN JOIN JOIN JOIN

Join the hundreds of students, faculty, news media, rock bands. Plus... a captive Dean Coleman.

Don't miss it!!

SAS-SDS

A flyer put out on the afternoon of April 23, right after the occupation of Hamilton Hall began, signed jointly by SDS and SAS. Note the primitive printing process: a stencil was "cut" using a typewriter then run out on a mimeograph machine; the heading and comments had to be cut by hand with a stylus, because the typewriter did not have large font. Many thousands of flyers were produced this way, with mimeographs running around the clock.

David Gilbert, my mentor, leading a mass strike meeting at Columbia, May 1, 1968, the day after the bust. Though he was by then a graduate student at the New School, downtown, he was highly respected at Columbia for his years as an undergraduate organizer.

Juan Gonzalez and me helping a female student onto the ledge
outside Grayson Kirk's office in Low Library.

A meeting in Fayerweather Hall, one of innumerable meetings in the occupied buildings on questions such as the amnesty demand. I have no idea why I was wearing a tie.

Exhausted New York City police occupying the campus after their early morning riot, April 30, 1968. They cleared the buildings by beating students, faculty, and onlookers as they arrested more than six hundred people.

A thousand students and strike supporters rally outside Columbia's Amsterdam Avenue gate, May 1, the day after the first bust. A police attack began shortly after this photograph was taken, in which one cop was seriously injured.

Student Afro-American Society (SAS) march sometime after the first bust, April 30. From left, SAS leaders Ray Brown Jr., Cicero Wilson (carrying sign), and Bill Sales.

The prototype for *Doonesbury*'s "Megaphone Mark" at a rally in front of Alma Mater after the first bust.

Ted Gold addressing a press conference at the Columbia strike, April 1968. A skilled and sensitive organizer on campus, two years later he would be dead in the Greenwich Village town house explosion.

My mother kissing me at a Columbia strike meeting in September 1968. My parents alternated between being proud of me and terrified for my life.

The cover of *Newsweek,* September 30, 1968, shows the confrontation at Columbia in which we unsuccessfully tried to shut down registration for the fall semester. The mass media broadcast the Columbia strike around the country and the world.

During the school year 1968 to 1969, I spoke at dozens of college campuses around the country about what had happened at Columbia. This is a talk on January 7, 1969, at the University of Kansas.

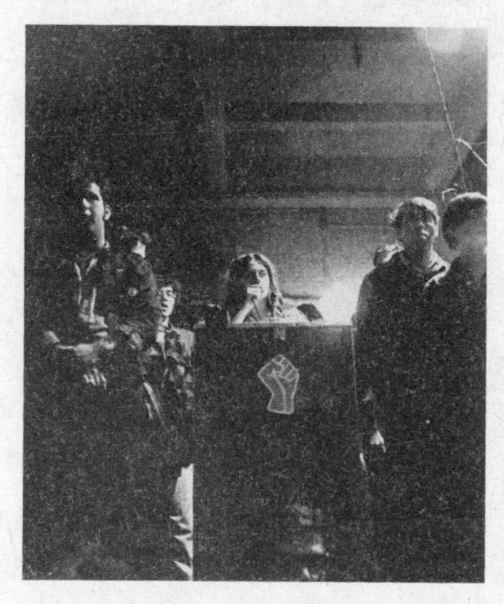

The last SDS national convention in June 1969 was a debacle involving a split between the "regulars," led by the Weatherman faction, and the Progressive Labor Party, a Maoist sect. Here Bernardine Dohrn dramatically announces the expulsion of PL. I'm standing just to her left.

On the last day of the Days of Rage in Chicago, October 1969, a line of determined Weathermen sets off to fight the police to "Bring the War Home!" I had just been jumped, beaten up, and arrested by plainclothes cops. From the left, Peter Clapp, my former roommate; John Jacobs (JJ); Bill Ayers; and Terry Robbins, who would die five months later in the town house explosion.

More than three hundred Weathermen were arrested and scores were injured in the Days of Rage. What we thought we were doing was building a "white fighting force," which would directly challenge state power, in the form of the Chicago Police. The whole thing was closer to suicide.

An FBI surveillance photograph taken outside the National War Council in Flint, Michigan, December 1969. I was by then dressing the part of outlaw.

On March 6, 1970, three of my comrades—Terry Robbins, Ted Gold, and Diana Oughton—died in this New York City town house, where the bombs they were making exploded prematurely. The intended target was a military dance at Fort Dix in New Jersey.

Prairie Fire was a full-length book published clandestinely in 1974 that put forth the Weather Underground's rosy view of impending world revolution, including inside the United States. It marked the turn from bombs to literature. In order to republish it aboveground, a network of hyperradical support groups known as Prairie Fire Organizing Committees came into existence.

After *Prairie Fire*, the Weather Underground moved toward becoming an illegal publishing house, which made no sense because the content was not illegal. In the spring of 1976, right before the organization dissolved, this last issue of *Osawatomie,* named after John Brown's Kansas battle against slavery, appeared. The cover photograph is the Pulitzer Prize–winning shot of white teenagers skewering a black man with a U.S. flag.

RIOT; CONSPIRACY

WANTED BY FBI

MARK WILLIAM RUDD

DESCRIPTION

CAUTION

IF YOU HAVE INFORMATION CONCERNING THIS PERSON, PLEASE CONTACT YOUR LOCAL FBI OFFICE.
TELEPHONE NUMBERS AND ADDRESSES OF ALL FBI OFFICES LISTED ON BACK.

These wanted posters appeared in post offices all over the country in 1970 and stayed up for a long time. I ripped down several, though.

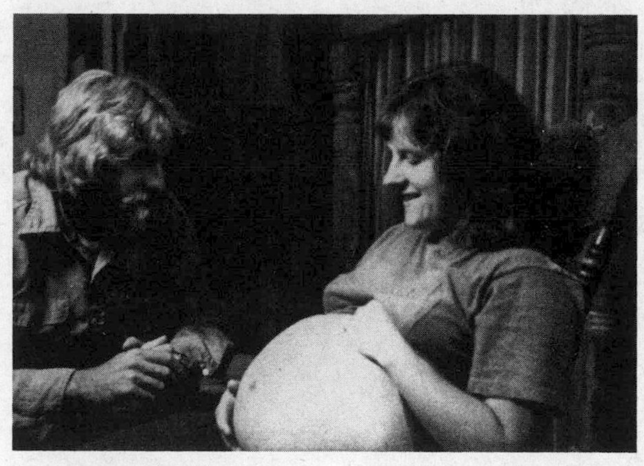

Sue, pregnant with Paul, and I admiring our prize watermelon, in our schoolhouse in Pennsylvania, summer 1974. This is one of the few pictures ever taken of us underground.

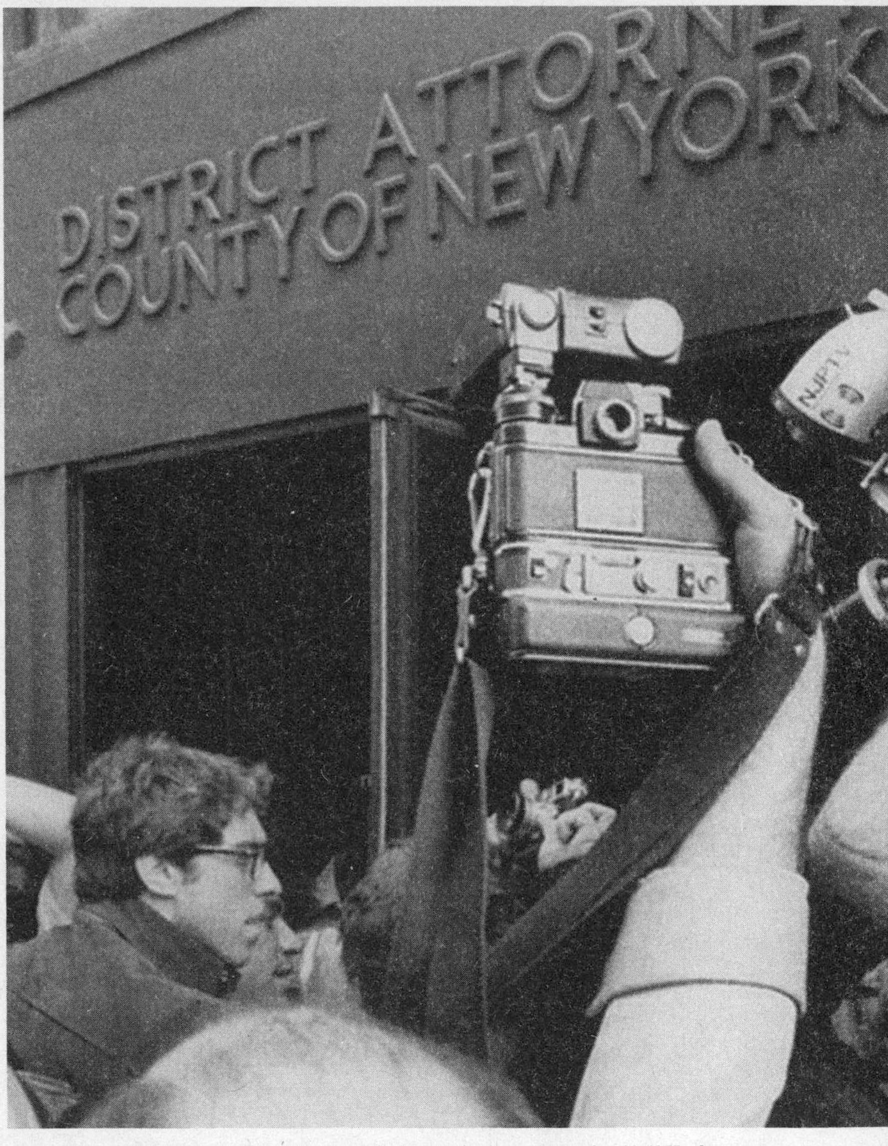

Turning myself in at the Tombs in New York City on September 14, 1977, after seven and a half years as a fugitive. My attorney and I had to fight our way through a mob of reporters. I was surprised at the extent of the press interest.

We landed in Albuquerque a year after I turned myself in. In this photo of our crunchy granola family, taken in 1979, I'm carrying Elena, one, and Paul, five. Sue and I look pretty happy, though we'd separate the next year.

This picture was taken in our backyard in Albuquerque, New Mexico, in the fall of 1978, a year after I surfaced. Despite the fact that we settled two thousand miles away from New Jersey, Jake and Bertha are pleased with not only their two grandchildren, but how the whole thing turned out.

Part of the crew of the New Mexico Construction Brigade to Nicaragua, near Estelí, in the northern region that was under attack by the U.S.-backed contra. That's me to the right, by the wall. Our goal was to show the Nicaraguans that not all Americans wanted to kill them and to come back to New Mexico with information to help build a movement to oppose the U.S. intervention.

Back at it, speaking to a history graduate students' conference in 2006 at Drew University, Madison, New Jersey. I told them about the difference between organizing to build a movement, such as the millions-strong anti–Vietnam War movement, and self-expression, which was where Weatherman and the Weather Underground went wrong.

For the rest of the meeting, we planned shifts of personnel around the country—closing down some collectives, strengthening others. In many cases we purposely split up monogamous couples and assigned the partners to different cities, in keeping with the anti-monogamy line of the fall. Of course Bernardine and her then partner, Jeff Jones, were exempt. Perhaps a hundred people were reassigned, much as military personnel are moved around by the top brass. Our job as leadership was to go out and explain the changes. I was given responsibility for the Cleveland, New York, and Denver collectives. I had convinced myself that shuffling the organization in order to prepare for the underground was what needed to be done.

I remember walking outside during breaks, feeling the relief of the warm Indian-summer sun. Why was I so tired, so overwhelmed with the feeling of being trapped? At night I'd withdraw into my sleeping bag, letting it shut out everything but the firmness of the floor. I wasn't willing to admit that I was in a state of shock and severe depression. Deep down I knew that the Days of Rage had been a terrible failure and that now we were about to make even more terrible mistakes. I felt like a member of the crew on a speeding train, dimly aware of disaster ahead but unable to put on the brakes.

On the third day of our meeting at White Pines, during a break, I noticed a large number of police cars down by the park headquarters. I quickly took two .38-caliber revolvers we had with us and buried them in the woods behind the cabin, thinking that we'd be worse off with them than without. A few minutes later, sheriffs and the state police burst into our cabin. They claimed that one of our cars, a rental vehicle, was stolen; it was one day overdue. They also caught Jeff Jones with a sap, a small blackjack, in his pocket.

Jeff just stood there with his arms handcuffed behind his back, wearing a wool Mexican poncho, a leather cowboy hat on his head, looking like Clint Eastwood in *The Good, the Bad and the Ugly.* So I lit his cigar for him, and we laughed.

Many similar harassment arrests, plus constant surveillance by the police and other agencies, caused us to turn inward to our small circles, the collectives. After the Days of Rage, each collective tried to prove its revolutionary worth, perhaps sensing that a move underground was coming. *Are you strong enough, brave enough, willing to be single-minded enough to be a revolutionary, a Weatherman? Would you give your child away? Would you sleep with me, or him, or her for the revolution?*

This short period in 1969, from mid-October to the end of December, was the most notorious Weatherman period—more group sex, LSD acid tests, orgiastic rock music, violent street actions, and constant criticism, self-criticism sessions. This is the time many people chillingly recall with tales of cat killing to prove our ruthlessness (I never saw it myself, but it could have happened) and of group orgies to prove our revolutionary love for each other.

We were by now a classic cult, true believers surrounded by a hostile world that we rejected and that rejected us in return. We had a holy faith, revolution, which could not be shaken, as well as a strategy to get there, the *foco* theory. Our logic was self-fulfilling: The failure of the National Action proved that we could rely only on ourselves. The more people left because they were fed up and unable to continue under the brutal collective and hierarchical system, the more our resolve was strengthened. The fact that no or few SDS chapters supported us proved the truth of our line that students were middle class and couldn't be trusted—with the exception only of ourselves (and even we were suspect). The rest of the movement hated us, which only confirmed the rightness of our path.

I spent my time traveling between the collectives to reorganize them and disseminate the latest correct line from the Weather Bureau. I would lead criticism sessions of the cadre, which would usually last late into the night, after which I'd bed down with my local girlfriend, who was often one of the collective's female leadership. I was also sick a good deal of the time, as were we all. Poor nutrition, fatigue, stress, and bad hygiene all contributed.

But there were also feelings of connectedness and relevance that I'm sure are common to all cults. We were revolutionaries, about to move to a "higher level" and begin armed struggle against the worst imperialist state in modern world history. We were the latest in a long line of revolutionaries from Mao to Fidel to Che to Ho Chi Minh, and the only white people prepared to engage in guerrilla warfare within the imperial homeland. It wasn't supposed to be easy.

Back in New Jersey, my parents were sick with worry, literally, during this period. Sometime in October, after the Days of Rage, my father had a series of small strokes, during which he'd lost his consciousness and memory for brief periods.

Often, in the months before, they had stayed up all night listening to the radio, trying to get news of my whereabouts. They heard me on the radio say that I was a Communist. This did not make them happy. Every time the phone rang, they expected some new disaster—a beating, an arrest; conversely, they waited for my occasional calls. After the Days of Rage, I didn't call them for about a week, even though they had sent sixteen hundred dollars to the National Office to help with bail. Not able to bear my silence any longer, they called the NO in Chicago looking for me, and the person who answered, an SDSer from Columbia whom they recognized, told them I was "in seclusion, but all right." They assumed I was being hunted by the police.

Back in September they had decided that I was off my rocker. Once, when they knew I'd be stopping at their house to borrow their car, they had waiting with them an old family friend, a physician who'd known me since I was a kid. They wanted Dr. Jacobs to evaluate me with the intention of having me committed to a mental hospital.

He started cross-examining me about my politics and state of mind, and I responded with the usual explanation about the need to build a revolutionary anti-imperialist army to help the people of the world. I tried to be as charming and loving as possible, stressing the humanistic concern at the core, but I could tell that my rap wasn't working. I didn't know at the time that this was an inquest into my sanity.

Finally, on an inspiration, I opened the briefcase I was carrying and said, "Look at this: Here's proof that we're not alone. I just picked this up in Cleveland from a supporter." Their eyes went wide. In the briefcase was ten thousand dollars in small bills, collected from an heir to some fortune or other who had contributed to the cause. The sight quieted them, and I was able to continue on my way, off to fight the revolution. The doctor didn't know what to make of it all. He said to my parents, "Maybe *we're* the crazy ones."

Jake's strokes, though, had me backed up against the wall. It was as if I were murdering my parents, they were so stricken with grief. But what should a real revolutionary be capable of sacrificing? I asked myself. Shouldn't I be willing to sacrifice *them,* even to hurt them, for the sake of the revolution? Around the world people were fighting and dying to defeat U.S. imperialism. Why should I be exempt and spare my family? Had Che worried about his parents when he picked up the gun in the Sierra Maestra or went he went to meet his end in Bolivia?

With a heart torn between my revolutionary duty and my guilt over my parents, I could only keep on going, putting foot in front of foot, toward my fate, dragging my parents with me.

11

To West Eleventh Street

The Giant Ballroom was a dilapidated dance hall located in the heart of the black ghetto of Flint, Michigan. On the morning of December 26, 1969, I examined a hole the size of a nickel in the plywood that had long ago replaced the glass in the front door. The night before, a disgruntled patron, bounced from the premises for reasons I never found out, had fired a single shotgun slug through the closed door. It had struck and killed an innocent Christmas reveler at the back of the hall. I pondered this whole business—the question of violence, random and otherwise—as I ran my finger around the hole: *what a fitting place for us to hold SDS's last National Council meeting.*

SDS had traditionally convened three meetings a year to act as interim decision-making bodies between the summer national conventions. Nominally, there was still an SDS: I was the national secretary; my office was on West Madison Street in Chicago; we still had a bank account and a printing press.

But in reality the Days of Rage had killed SDS. Earlier that fall an avalanche of chapters had disassociated themselves from the National Office. Others had folded up, their members demoralized by the factional fighting and violence of the past year. The few still attached to us were reduced to campus appendages of local off-campus Weatherman collectives. In June fifteen hundred had attended the previous national SDS gathering; attendance had now been reduced to around three hundred.

We decided to call this meeting the National *War* Council, but our pamphlets also called it a "Wargasm." The Detroit collective had decorated the ballroom unlike any other dance hall I'd ever seen. A six-foot cardboard machine gun suspended over the stage set the tone, as did psychedelic portraits of our heroes Fidel, Che, Ho Chi Minh, Lenin, Mao, Malcolm X, and Eldridge Cleaver of the Black Panthers. One wall was given over to alternating red and black posters of fallen Panther leader Fred Hampton, gunned down in his own bed just three weeks before. The Chicago police and FBI agents had stormed a Panther house a few blocks down West Madison from the SDS National Office and murdered Fred and another Panther, Mark Clark. Much of the black community of Chicago, plus every radical in the country, had reacted in anger; despite our pretensions to urban guerrilla warfare, we Weathermen had done nothing in response other than to send a delegation to the funeral. Now "Avenge Fred Hampton!" became our battle cry and obsession.

The five days of meetings resembled a rally more than a traditional SDS conference. Karate practice, singing of Weather songs written by Ted Gold and others ("I'm dreaming of a white riot, just like the one October eighth" and "Play, Elrod, play, play with your toes awhile" and "We all live in a Weatherman machine"), frenzied dancing to Sly & the Family Stone records with our own improvised words ("Che *vive,* viva Che!"), and speeches by Weather leaders replaced the traditional position papers, proposals, and resolutions. No votes were held.

The speeches were performances in themselves. I harked back to my high-school education: "We have to be like Captain Ahab, we have to become monomaniacal and take the harpoon of righteousness and kill the white whale of imperialism." Bernardine was more street: "We're about being crazy motherfuckers and scaring the shit out of honky America." And JJ, poetic as always: "We're against everything that's good and decent in honky America. We will burn and loot and destroy. We are the incubation of your mother's nightmare."

My own madness—possibly to keep up with that of my comrades—slipped out of my mouth as I paced the floor back and forth in front of the assembled troops. "It's a wonderful feeling to hit a pig. It must be a really wonderful feeling to kill a pig or blow up a building." Where did these words come from? Posturing alone doesn't tell the story. They came from my righteous anger—and my grief—over what our country was doing in Vietnam and what the police were doing here at home. My FBI files report me as also saying, "We are going to meet and map plans to avenge the deaths of Fred Hampton and Mark Clark."

There were crazy discussions at Flint over whether killing white babies was inherently revolutionary, since all white people are the enemy. Out of this bizarre thinking came Bernardine's infamous speech praising Charles Manson and his gang's murder of actress Sharon Tate, her unborn child, and the LaBiancas. "Dig it!" she exclaimed. "First they killed those pigs, then they ate dinner in the same room with them. They even shoved a fork into the victim's stomach! Wild!" We instantly adopted as Weather's salute four fingers held up in the air, invoking the fork left in Sharon Tate's belly. The message was that we shit on all your conventional values, you murderers of black revolutionaries and Vietnamese babies. There were no limits now to our politics of transgression.

After the speeches we set about breaking down the remaining collectives into cells, or affinity groups, or "tribes"—the terminology was never fixed. These small groups would begin the armed struggle: bombing police cars with crude devices made of gunpowder in aspirin bottles, doing anything to mess up the Machine. Also, the Weather

Bureau formally decided—in a closed meeting, of course—to abandon the SDS National Office and whatever remnants of SDS we still controlled. We felt we couldn't run an aboveground organization as well as a clandestine one. SDS was much too easily infiltrated by the police. The Flint War Council was infiltrated, we were sure, but we felt that the meeting was a necessary cost of making the transition to the underground, which would be secure.

A certain reality underlay our arguments: SDS was effectively dead anyway; after all, we had been instrumental in killing it. The National Office had ceased months ago to service the chapters; we had replaced *New Left Notes* with the Weather propaganda sheet, *FIRE!;* and, most of all, the vast majority of SDS members had turned against us.

Even the FBI had reached these conclusions about SDS. An FBI report I received through FOIA commented:

> The SDS has been fractured deeply in the last four months. Its pre-eminence as the leader of the young radical left in the USA is now questionable. By their stubborn adherence to pseudo-Marxist/Maoist dogma which is out of step with the present realities, RUDD and his colleagues have alienated a large segment of potential and heretofore willing followers.

I couldn't have said it any better. The anonymous FBI analyst and his superiors must have had a long laugh over the gift we'd handed them.

The destruction of SDS was probably the single greatest mistake I've made in my life (and I've made quite a few). It was a historical crime. The war was far from over—hundreds of thousands of U.S. troops were still in Vietnam, and we had yet to invade Cambodia and Laos. It would continue for five more years. We should have tried to use SDS to build as broad and powerful a movement to end the war as possible. Yet my friends and I chose to scuttle America's largest radical organization—with chapters on hundreds of campuses, a powerful national identity, and enormous growth potential—for a fantasy of revolutionary urban-guerrilla warfare. None of us in the Weather leadership, to my knowledge, were police agents either. We did it all ourselves. For decades I've been contemplating the wonder of this fact.

It can reasonably be argued that SDS was well on its way to oblivion by the summer of 1969—due to the factional fighting with PL and the gross acceptance of Marxist rhetoric. Maybe nothing could have saved it. But the fact is that my friends and I pushed SDS into the grave.

A few slightly saner people within Weatherman, intuiting that the leadership was wrong, tried to reverse the decision. Ted Gold, who had managed the National Office, argued for keeping the NO open and maintaining some presence on campuses, even as part of the organization went underground. Ted was an instinctive democrat. He felt that if our anti-imperialist line was real, we should take it out to students and educate and organize opposition to imperialism, not just retreat underground. I heard about these discussions with Ted secondhand, unfortunately, and did nothing to back him up. A few weeks later, in February 1970, Ted and I would load a VW van full of New York regional-office files and mailing lists and dump them onto a garbage barge at the sanitation department's pier on West Fourteenth Street. Ted was eventually won over to the total armed-struggle position: The following month he wound up at a Manhattan town house as a leader in a bomb-making unit.

As the new decade dawned, the New Red Army marched out from Flint, exhilarated and terrified. Despite my strutting performances at the War Council, I felt weaker and less sure than ever of my position as a general in that army. I was exhausted from playing a double role—the public revolutionary leader and the private scared kid.

I went to a woman friend's house in Ann Arbor on New Year's Day 1970 and dropped acid for the first time in my life, losing myself in sex, disorienting psychedelic sensations, and fearful paranoiac hallucinations of murder and death.

In January, at the next meeting of the Weather Bureau, we decided by mutual agreement that I would take a break from the leadership collective and join a clandestine "tribe" we were setting up in New York City. That would allow me to concentrate on just one task—building the guerrilla collective—while not trying to be the Big Leader of the whole organization anymore.

Since the late fall, I had been moving down in the hierarchy of the Weatherman organization, unable to "exert leadership." At my last Weather Bureau meeting, I initiated no plans for "submerging" the organization. Mostly I just sat and fumed about how Bernardine, Jeff Jones, Terry, and JJ, my former equals, were ignoring me and "seizing power." Half the time I'd blame them for my troubles; the other half I would blame my own weakness and inadequacy. I was so addled that I couldn't allow my conscious mind even a tiny doubt as to the direction of the organization.

By contrast, Terry and JJ, the two East Coast leaders, sure of where we were going, *were* providing leadership. In our many meetings in New York City, one or the other would rant, "White people are pigs. This whole society has to be brought down. We have got to defeat white-skin privilege; we can't let the Panthers and the Vietnamese bear all the costs." They were up, and I, by comparison, was down and feeling utterly weak. I was experiencing the competitive world of the Weatherman hierarchy from the underside now. How much worse it must have been the previous nine months for all the cadre beneath *me*.

JJ was in charge of my collective, or " national tribe," which was preparing to operate at a fully clandestine level, with untraceable safe houses and arms and munitions. Our goal was to be able to engage in armed actions, still to be determined, of a national significance. For now, living under our original names and identities, our task was to get together money and ID and the infrastructure of safe houses and cars and supporters.

Meanwhile, Terry Robbins was in charge of a separate "regional" collective of approximately a dozen people. They would not be as fully clandestine as the deeper group I was helping to build. But this "tribe" was still given the task of armed action in relation to local issues, such as the Panther 21 trial, then under way in New York City. Late in February the group tried and failed to ignite a firebomb at the home of the Panther 21 judge, John Murtagh. Terry, whom I saw regularly because of my privileged position as an ex-member of the Weather Bureau, told me he was ashamed of their incompetence.

On March 6, 1970, in the basement of a town house on West Eleventh Street in Greenwich Village, Manhattan, just a half block off Fifth Avenue, two wires crossed. This small accident completed an electrical circuit to a detonator cap, which in turn must have set off the dynamite. Neighbors reported hearing three separate blasts, which were undoubtedly the three homemade antipersonnel bombs—dynamite wrapped with nails—under construction. The timer consisted of a cheap pocketwatch with a screw drilled through the plastic crystal. In these primitive bombs, the hour hand was ripped off and the minute hand was to touch the screw and close the circuit from a dry-cell battery to the detonator less than an hour after the bomb was planted. Probably a tiny short had occurred while the timer was being assembled, something simple, preventable, and yet at the same time quite inevitable.

The blast rumbled upward from the basement, and then the four-story building collapsed in a roar of thunder as brick and wood and plaster dust fell onto three of my friends' bodies, killing them instantly, I hope.

Ted Gold, who had been the vice chairman of the Columbia University SDS chapter just two years before, had gone out to the corner drugstore to buy cotton to muffle the ticking of the watches. He came back to ask Terry Robbins, the group's undisputed leader, what kind of cotton to buy, balls or batts? *Why couldn't he make his own decisions?* I wondered in anger when I later heard the story. As he was on his way out again, the stone lintel over the basement door collapsed on him, one step away from survival.

Part of the problem was that Terry was doing all the thinking. He was just twenty-one years old, small, wiry, and smart as a whip, though by the time of the explosion his thinking had become twisted. A few nights before, Terry had told me what his group was planning. "We're going to kill the pigs at a dance at Fort Dix," he said. It was to be an antipersonnel bomb made out of stolen dynamite with sixteen-penny nails for shrapnel. Noncommissioned officers and their wives and dates in New Jersey would pay for the American crimes in Vietnam.

At that point we had determined that there were no innocent Americans, at least no white ones. They—we—all played some part in the atrocity of Vietnam, if only the passive roles of ignorance, acquiescence, and acceptance of privilege. Universally guilty, all Americans were legitimate targets for attack.

Terry's body wasn't identified until weeks later, and then only from a fingertip found in the ruin. Ted, twenty-three, and Diana Oughton, twenty-eight, were identified by the next day. Diana, tall and graceful, smart and earnest, had been driven in part by her experience as a Peace Corps volunteer in an impoverished Indian village in Guatemala. What she saw in Guatemala and learned about the war in Vietnam led her to despise our country's actions in the Third World.

I assented to the Fort Dix plan when Terry told me about it. I, too, wanted this country to have a taste of what it had been dishing out daily in Southeast Asia over the course of the previous decade. Our bombs would be crude, nothing like the sophisticated fifteen-thousand-pound "daisy cutters" used in Vietnam, the antipersonnel weapons with curlicued plastic shrapnel, diabolically designed to be undetectable by X-rays.

I spent most of March 6 at a friend's house in New Jersey, in part to establish an alibi in case anything went wrong with the Fort Dix action. Returning to New York, I called Robert Friedman, my old friend from the *Columbia Daily Spectator*, also to keep my alibi going, and by chance we decided to go see *Zabriskie Point,* the latest Antonioni movie. After the romantic young revolutionary hero dies, the film ends with the explosion of a fancy bourgeois house in the desert, a metaphoric fantasy of revolutionary retribution. Though I was uncritically enthusiastic about the movie when I saw it, I've never been able to watch it again.

I returned to my collective's house on Henry Street on the Lower East Side at about midnight. My comrades were huddled around the early edition of the next day's *New York Times.* "Where have you been? Didn't you hear?" they demanded, and then they showed me the front page.

TOWNHOUSE RAZED BY BLAST AND FIRE; MAN'S BODY FOUND, screamed the headline. A large picture of the burning rubble carried the caption "Smoke pouring from the four-story building at 18 West 11th Street, near Fifth Avenue. Explosions in the basement shattered glass in the area." The article mentioned that two women had escaped from the explosion, one naked, and were helped by a neighbor, the actor Henry Fonda's ex-wife, before disappearing. Later in the article, the paper ran another picture of the rubble and a shot of Dustin Hoffman, another resident on the block.

I had dropped Terry off at 18 West Eleventh Street two days before, though I hadn't gone inside.

A red gash split open in time. A stillness lasted less than a second, an eternity, before the pain rushed in. I was face-to-face with a loss so immense that it dwarfed everything else, yet I had to act. I willfully switched myself over to crisis mode—though at a deeper level I knew that there was now, suddenly, a before and an after—and in the second instant I was already thinking about what needed to be done.

First I had to find the survivors, to see who was alive and who had died, to try to regroup them and get them to safety.

I used a pay phone a few blocks from the apartment and miraculously, on my first call, managed to find a survivor. We had previously established an emergency backup number. I went to the apartment where she was being sheltered by a friend in the Village, only a few blocks from the Eleventh Street town house. The smell of smoke was still in the air.

We held on to each other as she told me the story. Terry and Diana had been working on the bombs in the basement. Kathy Boudin was taking a shower in a bathroom on an upstairs floor when the building blew up and then collapsed. She and Cathy Wilkerson, who was also upstairs, escaped together, screaming and coughing through the smoke and dust; miraculously, they were not injured. They were taken in by the neighbor, who gave her clothes; the two women cleaned themselves up as best they could, told the neighbor's maid they were going to a drugstore for first-aid supplies, then slipped away and separated. Since both women had been identified, and Cathy Wilkerson's father owned the house that had blown up, they would have to stay in hiding.

I found all the remaining survivors of the collective—including people who had not been in the house at the time—through more payphone numbers. We met the next morning at a Fourteenth Street coffee shop. Teddy was not among them, and no one had seen him. We wanted to believe he was safe somewhere, but we feared the worst. That afternoon the papers reported that Ted Gold, of Weatherman and Columbia SDS, was one of the dead, confirming our worries.

The survivors were subdued and in a state of shock. We could only focus on practical problems: where to find safe houses, how to get clean IDs, how to travel. Occasionally someone would say, "We just weren't trained well enough," or "It was our fault." If some of them were more critical, which they had to have been, they kept it to themselves. With three dead, no one dared question the basic strategy.

Like a zombie, I was going through the motions of continuing the underground. I remember thinking the cliché, *If you fall off a horse, the best thing to do is get back on.* (What did I know about horses?) I never asked if we should have been on the horse in the first place. Within a week I'd arranged for the surviving collective members to go to upstate New York for a day of shooting practice, presumably for some sort of therapeutic purpose.

Little by little, in talks with the survivors, I learned that the tension in the collective during the weeks before the explosion had been unbearable. Whenever anyone expressed a doubt about the planned Fort Dix bombing, Terry, Diana, or any one of the collective members would turn around with an attack: "You're just accepting your white-skin privilege," or "Don't you think white people are pigs?" It was the same "gut check" of the previous summer and fall, only raised to a higher, more dangerous level.

One collective member, an old friend from Columbia SDS, told me that Teddy had warned that anyone who pulled out of the action would have to be "offed" for the sake of security. I was stunned: This was completely uncharacteristic of gentle Teddy, who months before had argued for keeping the aboveground, legal SDS going and who at Columbia had always counseled caution. The town-house collective had spiraled into madness.

The collective had been structured hierarchically, a sort of military organization, with Terry at the top, followed by Diana and Teddy as his lieutenants. The remaining members were privates in the army, presumably subordinate because of less motivation, daring, or smarts. Slowly it began to dawn on me that we had created the perfect structure for failure. Part of the problem was passivity, especially blind acquiescence to leadership. If only each individual who now claimed opposition to the terror action at Fort Dix, myself included, had joined with the others and stood up to Terry. . . . But we had created a monster that suppressed dissent in order to expedite "what needs to be done."

Terry was so fierce that no one dared go up against him, even if we had seen things more clearly. He was a little guy, but with a temper and verbal adroitness that could draw blood. Many times over the previous year, he had turned on me for my weaknesses and doubts. In the months before the town house, Terry had traded his love of romantic folk music—I remember he got me to pay attention to the words of Leonard Cohen's "Suzanne"—for an obsession with two violent films that he watched repeatedly: *Butch Cassidy and the Sundance Kid* and

The Wild Bunch. In both these gory, sadistic movies, lovable charismatic outlaws are pursued into Latin America, then are massacred by local forces of law and order, but not before taking hordes of dark-skinned people with them. Not unaware of the ironies, in the Weather Bureau we called the endings "going out in a blaze of glory."

Terry had another violent aspect more deeply hidden: He regularly hit his girlfriend. Those of us in the leadership collective, the Weather Bureau, either denied that it was happening or knew about it and thought it wasn't important. I was among those who looked the other way. Today, if I were aware of a man brutalizing his partner, I would protect the woman without hesitation. The last thing I'd do would be to make the man the leader of an underground guerrilla group.

Terry had been playing out a terrible complex of violence, domination over women, and revolutionary heroism. I, too, used women for my egotistical purposes and extolled violence as a political act. I had made speech after speech telling people to "pick up the gun." But there was a big difference: As Terry's star rose in the organization—exactly when we were beginning armed struggle—mine was falling. At the time I could not possibly comprehend all the myriad underlying failures that came together in the town house; instead I suffered from a pain deep in my gut.

It would take me many years to figure out fully what had happened.

The survivors of the town-house collective were quickly scattered to other tribes around the country, on the theory that this was the best way to keep them from being busted. A few stayed underground, while most eventually drifted away from the organization and reestablished their lives within the aboveground radical movement. The most shattered dropped out of the movement entirely. I've completely forgotten who they are.

Much of that month after the explosion is similarly lost to my memory; it's easier that way. My pain at the time was a constant throb, though I attempted to shrug it off and continue my revolutionary work, now in the name not only of Che but also of our brand-new martyrs, Diana, Ted, and Terry. I never stopped to mourn, nor did anyone else I was with at the time.

Instead I went to work. I helped organize a plan to rebuild the organization in New York, which meant pulling together a functioning collective to look up old contacts and beg for safe houses and money; to rent apartments; acquire IDs, guns, cars with untraceable registrations—all the time-consuming and tedious housekeeping details.

Guerrilla warfare costs money, lots of it, so we tried an experiment in fund-raising. It was to be a simple operation, near closing time. Three well-dressed young white people in their early twenties, two women and a man, went into a steak house in suburban Westchester County, at a mall near a freeway. They didn't wait to be seated but instead more or less calmly showed the teenage-girl cashier their guns and demanded the money from the cash register.

Outside, I was nervously waiting at the wheel of the getaway car, which we had rented with a stolen credit card and ID. The minutes dragged on. We had estimated that the operation would take no more than four or five minutes total, yet it felt as if ten minutes or more had passed. Or was time just standing still? I was worried, thinking that something had gone wrong inside. In fact, the stickup had bogged down when our person assigned to clean out the cash register couldn't figure out how to open the machine—she had no experience in cashiering. Of course we hadn't thought of that problem. Fortunately, the manager, fearful of the guns and possibly taking pity on our incompetence, told the poor terrified girl behind the counter to open the register.

Finally the three came running out of the restaurant, jumped into the car, and we sped off the mile or so to our own car, made the switch, then drove cleanly back to Manhattan, with about twelve hundred dollars in cash.

We went to a fancy midtown restaurant to celebrate. Our steak and lobster were being paid for by the owner of the stolen credit card we'd used to rent the getaway car. Drunk on red wine, we began to express our doubts and mixed feelings. *How could we live like this for the rest of our lives? How long could we go on before we'd have to use the weapons to get our way? What if someone pulled a gun on us?* We had decided before the action that we wouldn't shoot, no matter what, but our victims didn't know that. How many mishaps would turn out as well as the cash register screwup?

It was obvious we had no future in armed robberies.

Shortly after that episode, the whole organization reverted to depending on our middle-class connections for money—family, friends, political supporters. Turning away from petty crime allowed us to concentrate on our main work, political bombings and building networks; it was also a hell of a lot safer to create aboveground support networks rather than rely on crime.

In early April 1970, we got news we'd been expecting out of Chicago. A federal grand jury, acting at the request of the Justice Department, had indicted twelve Weather leaders, myself at the top of the list, for the felony of conspiracy to incite riots during the previous fall's Days of Rage. Not only were we charged in *Rudd, et al.* under the very same law under which the Chicago 8 had been indicted the year before, but we were given the same miserable judge, Julius J. Hoffman! Our first action of the Days of Rage had been a rock-throwing march on Judge Hoffman's home at the Drake Hotel on Chicago's Gold Coast.

All twelve of the indicted had already disappeared as a result of the town-house blast the month before. Terry was dead, though still unidentified by the police. Kathy Boudin and Cathy Wilkerson were already wanted in conjunction with the town house, and the rest of us had been living as fugitives since March 6, not wanting to make ourselves available for the police. These new federal indictments brought no immediate changes to our lives other than that now we were official federal fugitives.

On April 15, not two weeks after the indictment, I was waiting in a coffee shop on East Twenty-third Street in Manhattan for Linda Evans, a new Weather Bureau member and a hard-core militant from the Michigan collective who had just arrived in town. Our meeting was set for 9:00 A.M. I arrived a few minutes early, got a table toward the back, and ordered a cup of coffee.

Just as the waitress brought the coffee, something clicked in my mind. On the way in, I had walked by several young white guys seated at the counter, all with medium-length to long hair and, most important, all wearing *brand-new* tie-dyed jeans (a hippie fashion fad of the time, in which portions of jeans were bleached). Somehow, miraculously, I had registered that tiny visual detail.

I got up, walked past the tie-dyed jeansmen to the cashier, put a dollar on the counter, and bolted out the door. Outside was another guy looking exactly like the ones in the coffee shop, only talking into a walkie-talkie.

This is it! I said to myself as I started running down Twenty-third Street with a pack of these guys in pursuit. I guess they didn't know for sure if I was their man, or else the sidewalk was too crowded with pedestrians, because they didn't shoot. At the subway entrance at Park Avenue, I ran down the steps. Had there been a train in the station, I would have jumped on, but since this movie didn't seem to have one, and I didn't want to get trapped in the station or on the tracks, I ran up another set of stairs, out onto the street again.

A city bus was loading right at the top of the stairs. I jumped on, threw some coins into the money counter (first rule of fugitive existence: Always carry change), and ran to the back of the bus. The bus was still stopped at the corner, and I was in plain view through the large windows. I threw myself onto the floor. People looked at me as if I were acting odd, even for New York, but of course no one said a word. As the bus began to pull away from the curb, I raised my head to the bottom of the window and peeked out. Not three feet from me, an agent was speaking into his walkie-talkie and looking about in all directions. I ducked down to the floor for a couple more blocks.

On automatic due to the adrenaline rush, I jumped off the bus, everyone on it looking at me, ran down into the subway, changed trains several times, then took a cab to the apartment we had on the Upper East Side. There I learned that Linda Evans had been busted that morning, as had Dionne Donghi, a cadre originally from Columbia, also wanted from the Chicago indictment, and Larry Grathwohl, who was traveling with Dionne. As we pieced everything together, Dionne had probably known about my planned meeting with Linda. She had probably told Larry, who we surmised had been an infiltrator, a government agent.

I had never met Grathwohl, but there had been suspicious reports on him from the Cincinnati collective, where he had joined Weatherman. A "greaser" and Vietnam vet recruited off the streets, he was a little too aggressive, a little too perfect in his hypermilitant line. Once, when the Cincinnati collective had administered an LSD acid test, Grathwohl had blurted out, "Yes, I am a pig," causing general freak-out in the room. Then he went on to explain that he was a pig because he had killed women and children in Vietnam. Having mouthed the correct line, the government agent passed the test and was welcomed into the collective.

Grathwohl had often proposed violence for street actions, claiming he was in Weatherman to act, not to talk. We liked to hear this kind of stuff. Since that time I've reflected on the fact that violent groups are the easiest for agents to infiltrate, because their line is so crude and easy for cops to understand. Still, Grathwohl was the only known infiltrator ever exposed in the Weatherman organization. Had there been others, the government certainly would have used them to bust us in the years after 1970. In a troubling and self-serving 1976 book, *Bringing Down America,* Grathwohl claims he tried in vain to convince his government handlers to delay the bust until he had information on more fugitives than just Linda and Dionne. He was working directly under Guy Goodwin, a Justice Department bigwig who had responsibility for our cases.

Grathwohl was right: If he had waited a day or two, he might have been able to bust me and several other Weather leaders. John Mitchell's Justice Department must have been under such intense pressure from the White House to get arrests within a few days of the indictment that they blew their sole infiltrator. The Nixon administration took quite seriously the revolutionary threat posed by the Weathermen. They were our truest believers—after ourselves, of course.

12

Mendocino

Following my narrow escape from the feds on April 15, I thought it would be wise to get out of town. Manhattan offers excellent anonymity because of its size and congestion, but I was becoming way too nervous to operate there. The cops now knew I was in New York. I spent my time shuttling between a tiny apartment on East Eighty-second Street and the walk-up on Henry Street on the Lower East Side, and also going to countless enervating meetings with "contacts"—potential aboveground supporters. Some of these restaurant meetings paid off with resources such as money or places to stay, but many didn't.

I was a lame duck in the New York organization. In a month I was slated to go to the West Coast for the first post-townhouse Weather Bureau meeting. Although I was officially off the Weather Bureau by then, I was being included in this important meeting "for historical reasons."

Philadelphia was a logical place to hide out for the next

month, since it was large and anonymous like New York and also close enough to the city for me to keep in contact with the Weatherpeople still there. I had never been there before but quickly became intrigued by its old downtown and its neighborhoods and parks. I spent days exploring, losing myself among the colonial alleys of the old parts of the city and the museums and galleries on the Parkway. It was a relief to be out of New York.

A New York contact had given me the name of a couple I could approach for a place to stay. In effect, I would be asking people I didn't know, friends of friends, to harbor me, a well-known federal fugitive, at great potential risk to themselves. Barbara and Carl were both in their late twenties, attractive artists living in a small one-bedroom apartment in a converted town house not far from trendy Rittenhouse Square. They were smart, productive people who led normal, uneventful lives. In walked the Weatherman.

They immediately agreed to let me stay on their couch, no questions asked. Like so many other people, they hated the war. They were not activists, however; I'm not sure either had ever even been to a peace march. Still, they knew I was a fugitive and were willing to help. We discussed keeping my comings and goings to a minimum so as not to arouse the suspicion of other tenants in the building.

In the course of the next week, we found we liked each other. Carl and Barbara brought an air of domesticity to my life, which I desperately needed: a pot of brown rice sitting on the stove, vegetable dinners around their tiny table. These decent, gentle people provided a soothing antidote to the tense, "correct" attitudes of the comrades from whom I had just escaped. Being with them helped salve my townhouse wounds.

For my part, I seemed to add just the needed touch of romance to their lives. We'd talk long into the night about politics and culture and sex and relationships. I was honest about what I believed, from the need to fight imperialism to smashing monogamy. I must have seemed like an exotic visitor from another country.

I stayed in touch with the organization via pay-phone-to-pay-phone calls, at prearranged times. I'd often sit in an old drugstore waiting for calls that were an hour or a day late. I passed the time reading newspapers, wondering in a funk whether this was how I'd spend the rest of my life. One day I got a message that JJ was coming down to Philly. *Finally, someone from the organization to talk to.* I arranged with Barbara to put him up at her art studio. Although the landlord didn't want anyone to stay overnight, Barbara was willing to take the risk for us. And this despite the fact that she tended toward pacifism and wasn't in agreement with the Weather line of armed struggle at all.

I'd always loved JJ's high energy and his raps. In many ways he had been my teacher over the years since we first met as freshmen at Columbia. I had learned internationalism from him—it was he who'd first told me of the national-liberation movements sweeping the Third World. His daring during the Columbia strike had challenged me to keep going, despite my periods of confusion and demoralization. Now he was my leadership in the underground; I was eager to hear his ideas about what direction the organization should take.

We weren't together but a few minutes before JJ announced to me that we should be putting together specially trained squads to engage in "higher-level" actions—squads that would actually create material damage to the war effort, like blowing up a B-52 on the ground or knocking out a vital government computer. He thought we might even consider a selective assassination or kidnapping.

My first reaction was fear. More people would certainly die from these actions. But in my befuddlement, his arguments quickly won me over: It's futile to build a mass movement to end the war; since we're really representatives of the Vietnamese, we have to fight in the same way as the Vietnamese. I agreed to back him up at the coming Weather Bureau meeting on the West Coast.

One afternoon I arrived at Barbara's studio to meet JJ and found him and Barbara curled up naked together on a mattress. I was a little unnerved and upset, worried that this new development would mean trouble for us—jealous husbands can become terrible security problems.

However, I sat down on the mattress, and in a few minutes Barbara and I were stroking each other and making out. JJ was asleep—or pretending to be. Barbara said she had been attracted to me from the first, and that she was bored in her marriage. JJ, with his strong arms and baby blue eyes, had come along as an opportunity she couldn't pass up. Now I was here, and it seemed to her like a great chance to continue the adventure.

Barbara was attractive and bright and attentive. Many times in the previous weeks, I had fantasized about her, but we had consciously maintained a controlled flirtation. Ever the sexual opportunist, I jumped at my chance now. So, quietly and in our respective fantasy worlds, we made love with JJ lying next to us, either sleeping or pretending to.

This wasn't the first time JJ and I had shared a girlfriend. Once, in college, a woman had told me that while she was making love with JJ, she was thinking of me. I'm sure, from the comments of other women, that the reverse had sometimes been true, too. The one time, a year earlier, when we had tried a threesome, I had felt an intense excitement at the thought that my semen was mixing with JJ's inside a woman. Now, once again, I was experiencing that same feeling, bonding with JJ through a woman.

It wasn't only JJ that I was thinking of as I lay with Barbara—there was also myself. In my mind I was Barbara's liberator, bringing her freedom from her life of quiet desperation. My penis was a magic wand of liberation. Sex was also, for me, a respite from my lonely days and nights contemplating the deaths of my friends and my own dismal future. At times of emotional crisis and depression, I always needed a woman.

God knows what was going on in Barbara's head as she wrapped her small, firm body around her second revolutionary that afternoon.

That night when I entered Carl and Barbara's apartment, they were in the midst of an argument. My fears had come true: Barbara had immediately told Carl about that afternoon. I was panicked but attempted to console Carl by telling him that women's liberation shouldn't threaten him. I also told him not to be jealous, saying something like, "Barbara's attraction toward JJ and me shouldn't reflect on your relationship."

Carl looked at me, stunned. "What do you mean, 'JJ and you'?"

I instantly realized that Barbara hadn't told him about me, only about JJ! She knew that hearing about his wife fucking two guys in the same afternoon would be too much for Carl to take. I was able to backtrack and weasel out of my faux pas by saying, "I only meant her *feelings* of attraction; we didn't do anything."

JJ left Philly immediately, but I stayed with Carl and Barbara a tense additional week, until it was time to go to California. I did get together once more with Barbara. She liked her new self, she told me; she felt free of the old repressed person.

I suppose this justified the whole affair in my mind. Carl was sullen, resentful, probably too shattered to throw me out of his house. The slash-and-burn, scorched-earth policy of the Weathermen had left destruction in its wake yet again.

After I left, I never saw Carl or Barbara again.

On April 30, 1970, acting solely by presidential fiat, Nixon sent U.S. troops, backed by B-52s, into Cambodia to hit the enemy "headquarters." Many people had been lulled into thinking the government was diminishing the American role in the war through troop withdrawals and "Vietnamization," which meant turning the actual ground combat over to the puppet army we had created, the Army of the Republic of Vietnam, or ARVN. Ten days earlier, on April 20, 1970, Nixon had announced a troop withdrawal of 150,000 GIs within the next twelve months. But Pentagon and administration war planners assumed that a "Central Office, South Vietnam," or COSVN, *must* exist, as a mirror image of our own Pentagon West, the U.S. military command center in Saigon. In actuality, U.S. planes had been secretly bombing Cambodia (a secret only to the American press and public) since at least March 1969, and U.S. soldiers had often made incursions across the border. No sign of a COSVN was ever found.

In justifying widening the war to a neutral country, Nixon used one of his more enduring and ironically appropriate phrases, saying that the United States would not accept defeat in Vietnam and act like "a pitiful, helpless giant."

Within hours of the invasion of Cambodia, college students spontaneously went into the streets with protest demonstrations. That weekend saw riots at Stanford and Ohio State University, among others, and at Kent State University in northern Ohio the ROTC building was burned down. Student-newspaper editors met on Sunday, May 3, to jointly denounce the invasion and call for an immediate nationwide student strike.

The next day, Monday, May 4, the Ohio National Guard fired on protesting students at Kent State, killing four and injuring nine. This was shocking for many white people, especially those who had not anticipated such violence from the government. (Ten days later two black students would be shot dead by police in Jackson, Mississippi, receiving little press attention.) Over the next week, 4 million students participated in a national student strike, making it the largest student action in world history. Five hundred campuses were closed down, while hundreds of thousands of people demonstrated in the streets. The National Guard was activated on campuses in sixteen states.

Meanwhile, I was sitting on a park bench in Philadelphia's Rittenhouse Square, thrilled as I read about all this in the *New York Times.* Finally the campuses were erupting en masse against the war, and it was bound to have an impact on public consciousness and government war policy. Yet here I was, hiding out, completely cut off from the protests. I didn't even dare walk the few blocks to join the demonstrations at the University of Pennsylvania campus for fear of being identified and arrested by police or federal agents watching the crowd.

Just months before, my friends and I in Weatherman had closed the SDS National Office, which could have served to coordinate the national strike and at the same time push our anti-imperialist politics. Terry, Diana, and Ted, all excellent student organizers, were now dead, and the rest of us were unable to function in the mass movement.

It was not widely known that Kent State, a school that drew from the sons and daughters of Ohio tire and auto workers, had been one of Weatherman's bases. Terry Robbins had organized there for years. The university's extremely militant SDS chapter had been banned from the campus in the fall of 1969, with Terry and the entire chapter leadership having been jailed for as long as six months. The chapter had produced dozens of Weather cadre, who had organized a collective in nearby Akron the previous summer. I had visited the campus several times, the first being a speech to an enthusiastic audience of about a thousand students in the fall of 1968. In some measure, the militancy of the university's Cambodia demonstrations resulted from confrontational politics that Weatherman had helped create at Kent.

Reduced to only reading about the mass student protests gripping the country, I was feeling nothing but isolated and powerless. Plus, I was still in shock from the loss of my three friends. Unable to acknowledge the emotions overwhelming me, I had acquiesced to JJ, promising to back him up in his proposal for more armed action at the upcoming Weather Bureau meeting. My mind was a tangled ball of confusion.

Before leaving Philadelphia for California, I took care of one final piece of business: I arranged to meet with my parents. I had not seen or talked with them since before the town house. Back on March 6, I had guessed they were panic-stricken at the possibility that I was one of those killed. WBAI, the leftist, listener-sponsored radio station in New York, had even reported a rumor that I'd died in the explosion. Through an intermediary I had sent my parents a message that I was alive, but they were still terrified for me now that I was a fugitive.

In Philadelphia, at a Center City delicatessen, I met my cousin

Miles, a young professor at Temple University. He agreed immediately, out of family loyalty, to bring my parents to a meeting in New Jersey.

Our reunion occurred on a rainy and cold spring day. I had made the mistake of renting a room at a cheap, dingy shore motel, a place I remembered passing as a child on the way to the beach. The room had paper-thin walls and leaks in the bathroom plumbing. My parents were restrained. They didn't carry on too much—what could they say?—though my mother did cry. My father was grim and tight-lipped. They told me how scared they'd been in the days before they received the message that I was alive. My father mournfully remembered a call from Terry Robbins's father, wondering if they knew anything. Mr. Robbins, whom my father didn't know, was probably fishing to find out if perhaps I, and not his son, was the unidentified third body. My father was deeply shaken by the call. "Why should Robbins's son, or anyone's son, have died?" he demanded to know.

I went through the litany of justifications for what we were doing: the war, imperialism, racism, the system of violence for which the U.S. government was responsible; the need to change the system; the growing revolutionary movement. By now this was such an old argument between us that it seemed useless to pursue it. They were terrified that they'd never see me again.

I was torn apart. I was so sure I knew better than my parents; after all, their generation had brought the world to this state of affairs, if only by their acquiescence. On the other hand, I so deeply loved these people who loved and adored me. I was paying them back for all the years of care and sacrifice—for the rides to music lessons and Hebrew school, for the pots of spaghetti and meatballs they brought to my apartment at Columbia—by completely turning my back on them. I *knew* that I might never see them again.

As we parted, after less than an hour, my mother pressed a roll of cash into my hands. "Better we give you this than you get arrested dealing dope or even worse," she said.

After they left, I just sat on the bed in that miserable room and cried.

I used the money to fly out to San Francisco later in May, going half-fare standby with a forged ID. The Bay Area was the organization's new center, not because it had been a focus of great Weather activity—we'd never even had a Weather collective there—but because it was the epicenter of the youth counterculture. The West Coast was conducive to our underground survival. Northern California provided city and country communes, student ghettos and street scenes, a whole panoply of possibilities that would give us camouflage and safe harbor.

California was also the home of the Black Panthers and the prison revolutionaries, including George Jackson. It was the jumping-off point for the thousands of GIs still rotating through Vietnam and the thousands more returning from the war; it was the most militarized of all the states, and in many ways the most repressive, so it seemed logical that we should have a presence there.

An old comrade, a member of our San Francisco tribe, met me at the bus from the airport. He had long hair, a beaded necklace, bell-bottoms, even a bandanna with a peace sign on it. By comparison, my dyed-black short hair and slacks and dress shirt made me stick out, hopelessly straight and East Coast. He took me to an apartment on Pine Street, on the Tenderloin side of Nob Hill, where I was reunited with Weatherman friends I had not seen since Flint. I felt as if I'd come in from the cold.

The next day the leadership showed up: Bernardine, with both her current consort, Jeff Jones, and Bill Ayers, who was soon to take his place. They didn't live with the troops at the tribe house, of course, nor would they say where they lived, for security reasons. (I later found out it was on a houseboat in Sausalito.) Bill and Jeff were dressed in their version of youth-culture attire—leather vests with beads, leather hats, jeans, and boots—while Bernardine had on a see-through net blouse, Grace Slick style, under her beads. I wondered whether the blouse made for better or worse security.

We embraced as slightly estranged brothers and sisters who had suffered a loss in the family. I was told we were leaving for Mendocino right away and that the meeting would start the next day. I was also told that JJ had arrived earlier and that people thought his armed-squads idea was atrocious.

About a dozen of the most wanted political fugitives in the country, all old friends, sat in the living room of a rented beach house. There was a big patio door facing the ocean. Bernardine started off the meeting with a review of the events leading to the town house. She had done her homework.

The New York collective, led by Terry, had begun its work in February with a firebombing of the Panther 21 judge's home. Since there had been almost no damage, Terry was in a frenzy to "up the ante." The collective's leadership had been sucked into Terry's mania for a big bombing, so they chose the Fort Dix target. Everyone in the collective was so frantic, poorly trained, and inexperienced, that they had not even designed a safety switch into the circuit—a simple safety switch. In fact, they were leading double lives, both above- and belowground, and many trails to aboveground contacts were left behind for the FBI, especially through telephone records. Everything had been done in a hurry and was very slipshod.

The Fort Dix action, according to Bernardine, speaking for the West Coast leadership, had represented the worst of Terry and JJ's politics, that all Americans are the enemy. I was struck immediately by how right Bernardine was. A terror attack on an NCO dance, most likely killing innocent people, was far off the political target. If successful, it would have had the effect of turning millions against us. And it was true that Terry and JJ had planned it. I suddenly realized how stupid I was to have been won over to JJ's current plan to up the military damage by creating heavily armed squads; it was just more of the same. Much worse, I also felt deeply guilty that I'd assented to Terry's Fort Dix plan, which I'd known about, too. Had I raised objections at the time, perhaps I could have prevented the tragedy.

Bernardine's opening set the tone for the whole week's discussion.

We would now try to develop a "new line," according to the West Coast leadership, learning from our mistakes and moving on. The organization had to be more "life-affirming," embodying the lessons of the counterculture. We had to acknowledge the importance of the mass movement—we'd all been struck with the size and breadth of the demonstrations after Kent State and Cambodia—instead of emphasizing the armed actions of small bands of urban guerrillas.

Still, the underground's function was to continue with armed actions, though not targeting people anymore. We would call in warnings before blowing up government offices, for example. We would pick only targets that would symbolize the underlying violence of the government and corporations; our term for this was "armed propaganda." The act of bombing buildings, an extreme form of resistance, would surely push the mass movement toward greater militancy, but without hurting anyone and having us labeled as terrorists.

After a few days, Bernardine announced that JJ would have to leave the organization. He had to go out on his own, she said, to learn about the emerging youth culture and "to get his head straight." She also confirmed my demotion in the organization: I would be brought into the San Francisco tribe as a cadre in order to reeducate myself about the youth culture. I had been too close to JJ and was completely expelled from the leadership.

That night JJ and I went out to a bar in Fort Bragg, a slightly larger working-class town up the coast from Mendocino. We drank and played pool. In the background Creedence Clearwater Revival played on the jukebox: "Looks like we're in for nasty weather." JJ agreed he had to leave the group. "I'm accepting my expulsion for the good of the organization," he told me. "Someone has to take the blame. Bernardine, Billy, and Jeff are right about the military error."

"But everyone knew what was being planned," I said. "We were all together in New York with Terry the week before the action, and nobody raised any objections."

"It doesn't matter. We have to create the fiction that they were always right so that they can lead the organization," he replied.

JJ seemed to me like a victim of one of Stalin's purges ready to falsely confess, for the good of the party, to being an agent of the imperialists. We talked about one of his favorite novels, Arthur Koestler's *Darkness at Noon*.

"I always respected the fact that the old Bolshevik confessed for the sake of the revolution," he told me, "there had to be a single unified revolutionary party, even under Stalin's leadership. The individual doesn't count; it's only the party and its place in history that's important.

"At least they're not going to liquidate me," he said with a laugh.

I was sad for JJ, but I also agreed with the criticism of his "militarism." Plus, I had gotten off easy: At least I wasn't being cast out.

"I'll be back," he assured me.

For the new leadership, JJ's expulsion was a brilliant maneuver that successfully rewrote history. Suddenly no one remembered how universally accepted the old "Fight the people, all white people are guilty" line was. The new regime regarded that as JJ and Terry's error and no one else's. Weather's history had been conveniently cleansed. And, in a double whammy, the strategy of bombings would continue despite the new leadership's embrace of mass youth culture. Targeting only buildings, not people, we switched over to "bombing lite." Not targeting people was morally an improvement, but politically it wasn't much of a change, because it continued the overall strategy of clandestine armed struggle.

As if to confirm that nothing had fundamentally changed, a "Declaration of War" was drawn up as the first communiqué from the Weatherman Underground, sent through the mail to underground newspapers and movement radio stations throughout the country. It was suitably contradictory, first extolling the mass struggle against U.S. imperialism but then claiming that our task was "to lead white kids into armed revolution," appropriating the guerrilla strategy of the Vietcong and the Tupamaros of Uruguay for "our own situation here in the most technically advanced country in the world."

The communiqué confirmed Terry as the third person killed in the town house, since his body had been so mangled as to have been questionably identified. Then it moved on quickly to proclaim our unity with the youth culture: "Dope is one of our weapons. . . . Freaks are revolutionaries, and revolutionaries are freaks. . . . If you want to find us, this is where we are. In every tribe, commune, dormitory, farmhouse, barracks and townhouse where kids are making love, smoking dope and loading guns—fugitives from American justice are free to go."

As if that taunt weren't enough, the communiqué ended with the promise that we would attack a symbol of "Amerikan injustice" within fourteen days as our way of celebrating the black revolutionaries who "first inspired us." "Never again will they fight alone," the communiqué ended.

Eighteen days later, on June 9, a bomb was set off inside the New York City police headquarters. Due to a phoned-in warning, no one was seriously injured. A statement entitled "The Second Communiqué from the Weatherman Underground" was received by the press the next day. It said, "The pigs in this country are our enemies," citing the death of Fred Hampton; the murders of six black people in Augusta, Georgia, and two at Jackson State, Mississippi; the four at Kent State; the imprisonment of Los Siete de la Raza in San Francisco and continual police brutality.

This statement concluded by expanding on Mao's famous phrase, "Political power grows out of a gun, a Molotov, a riot, a commune, . . . and from the soul of the people." Thus did the Weather Underground mourn the deaths of our three comrades and defeat "the military error."

PART III

Underground

(1970–1977)

13

The Bell Jar

The main work of the San Francisco tribe was survival, with an occasional bomb blast. In our heads there existed an ideal of unity, friendship, and intimacy associated with collective living and work. The reality was that the six or so of us trapped in the tribe rated ourselves by how close we were, socially and politically, to the three top people, Bernardine, Billy, and Jeff.

I was at the very bottom, because they wanted to have as little to do with me as possible: I was too unhappy and cynical, and I reminded them of JJ. My old friends, a few going back to Columbia, like David Gilbert, who had organized me into SDS, kept me at arm's length. Actually, we all kept each other at a distance. In that claustrophobic apartment on Pine Street, we played the Temptations' "I Can't Get Next to You" again and again, until it became the tribal theme song.

I spent as much time as I could away from the house. I hung out on Fisherman's Wharf, listening to the drumming on the steps by the Maritime Museum, looking out over the bay and the Golden Gate Bridge, Otis Redding's melancholy crossover hit, "The Dock of the Bay" ringing in my head: "Sittin' here resting my bones / and this loneliness won't leave me alone." Exploring the city on foot, up and down the hills, its parks, its shoreline, its Victorian houses—a living architectural museum—I appreciated every detail, every nook and cranny. The city was my grand distraction.

I hitchhiked along the coast getting to know the countercultural scenes in Sonoma, Marin, and Santa Cruz counties, hanging out at coffeehouses and food co-ops, meeting people and starting conversations. Occasionally I'd have a good conversation with somebody, about some random topic or other, and we'd get to know each other a bit. I'd eventually have to tell some vague story about dropping out of school in the East. Amazingly few people asked questions, probably because my story wasn't all that different from those of the people I was hanging out with—ex-students, ex-flight attendants, ex-secretaries, ex-everybody who had found acid or some other truth which led them to drop out of straight society.

I heard that the Scalers and Painters Local, a black-run dockworkers' union, occasionally hired casual laborers to help unload ships when all their regular members were out on jobs. You'd have to buy a union membership card, then hang around the hiring hall down by the Southern Pacific station twice a day—at 6:00 A.M. and again at 4:00 P.M.—for shape-ups. It was worth the trouble, since the pay was good, even if you worked only once or twice a week, which was my average. Because of the strength of the longshoremen's union, I'd typically make $60 to $80 a shift, which in those days was excellent money, the minimum wage at the time being $1.90 per hour.

Several times I was sent out on crews to load containers onto ships bound for Southeast Asia. The containers were sealed, so I never knew what we were loading, but I always assumed the worst—that we were sending some sort of war matériel to help murder people in Vietnam. Other than that, I liked the work and the guys I hung out with in the hiring hall and over lunch, most of whom were lifelong union dockworkers, predominantly black. They were very aware of the class nature of society and mostly all in opposition to the war. The paradox, which I couldn't articulate to them for fear of drawing attention to myself, was that we were probably making our living off the war.

One foggy night shift, I was on a crew at the Port of Oakland docks loading sealed containers for Thailand and Vietnam. My job was to hook the crane cable onto the containers being lifted off flatbed trailers and then unhook the crane from the containers being dropped onto the trailers. Hook, unhook, hook, unhook, all night long.

After I caught on to the work and got into the rhythm, my attention drifted off to my own secret thoughts—I can't recall what about, but it could have been anything. Lost in my reverie, or perhaps brooding, I hooked when I should have unhooked. Suddenly the trailer rose up into the air with the container, its wheels dangling over my head, and I jumped back about twenty feet, probably the most athletic maneuver of my life. The crane operator realized the problem a second later—since everyone around was screaming—and abruptly dropped the mess with a solid crash of metal. It was a tough way to learn to pay attention.

During those first six months in San Francisco, no one ever asked me more about myself or my background than I offered; there seemed to be a general understanding in that culture not to ask too much. Perhaps everyone had something to hide, emotional and/or legal. Also, I looked like so many others—a hippie dropout in his early twenties, military jacket, with hair getting longer every day—that there was no reason to scrutinize me beyond my disguise and phony ID papers

Still, the worst part of those months—worse even than the fear of getting caught—was the loneliness. I was nobody, doing nothing except surviving, with no real bonds to anyone. It's not a good way to live. *Will this be my life forever?* I found myself thinking.

Around the middle of the summer, I heard from a contact that my old girlfriend from Columbia, Sue LeGrand, was living in Oakland.

Sue and I had been together at Columbia for a year, from the summer of 1967, the Golden Summer of *Sgt. Pepper's Lonely Hearts Club Band,* through the events of the Columbia strike, to the following summer when we broke up, mostly because of my infidelities. Seeing her was a simple matter of getting an intermediary to arrange a meeting in San Francisco.

Sue's participation in the strike caused her to be one course short of graduating Barnard in the summer of 1968, but she took off for Europe anyway. There she worked, bummed around, and had her own adventures. At the end of 1969, after the Days of Rage, she came back to the States, feeling she was missing out on something important. She was one of the few people to join the Weathermen at Flint and participated in the death throes of the aboveground organization, in the Denver and Seattle collectives. After the town house, most of the smaller, outlying collectives were abandoned by the central organization, leaving the members to fend for themselves. Sue wound up in the Bay Area.

We met in Tommy's Joynt, an old San Francisco landmark at Geary Street and Van Ness, and were overjoyed to see each other. Sue had been worried about me since the town house and was overjoyed that I was okay, or at least alive. We began catching up on the lost time since Flint: me telling her about the town house and the aftermath, my life as a fugitive, she filling me in on the end of the Denver and Seattle collectives and on her new life since then.

We left the bar and were walking down a street in downtown San Francisco, holding on to each other, and an old lady stopped us, wagging her finger, "Don't do that out in the street. Go home and go to bed." Unfortunately, it wasn't that easy.

Sue was living an aboveground life in Oakland; she had a full-time job as a photo finisher at a Fox Photo lab. I was a well-known federal fugitive, living in clandestine safe houses with other fugitives. We had to assume that Sue appeared in my FBI file as my girlfriend at Columbia. It was only a matter of time before the feds, in hopes of finding me, would get to her.

So we met secretly, on weekends, in San Francisco. Occasionally we'd hitchhike out of town for the day or overnight, but mostly we'd spend our nights at a cheap hotel in Chinatown or the Tenderloin and our days wandering around the city. It would usually cost us no more than twenty dollars for the whole weekend—food, movie, hotel—money one or the other of us would scrape together that week. For me those dates were grand escapes from my grim tribal existence.

I felt trapped in the organization, always worried I would say the wrong thing, fearful that I'd expose myself as not being revolutionary enough or now, with the new line, not hip enough. The "collective" barely worked together.

Thanksgiving morning in 1970, for example, I suddenly realized that the tribe hadn't planned any celebration. I was feeling holiday loneliness for my family: I guessed that my folks in New Jersey were probably as miserable as I was. I set out about noon to find a turkey. After several hours I finally located one at the only open grocery store in all of San Francisco, somewhere out by the zoo. I prepared it as quickly as I could, given that the turkey was frozen. At about eleven that night, as a fitting cap to our dismal holiday, my tribal comrades and I sat down to eat a mostly raw bird.

But the weekends spent with Sue were pure delight. Finally I had a friend, a companion, someone to talk with about the problems of life, someone to love. Sue understood my predicament and began referring to the organization as "the Bell Jar," after Sylvia Plath's novel of claustrophobic madness.

Nineteen seventy was the year of the prison movement. Black and Latino prisoners around the country, and especially in California, had begun to redefine themselves as political prisoners. This process had been under way since the days of Malcolm X and the Black Muslims, an aspect of the growing anticolonial and nationalist consciousness of nonwhite people in the United States. In prisons around the country, convicts were educating and organizing themselves, resisting the conditions of their incarceration, demanding basic civil and human rights such as decent food and an end to solitary confinement and other arbitrary punishment.

Eloquent books by black prisoners, such as Eldridge Cleaver's *Soul on Ice* and George Jackson's *Soledad Brother,* became bestsellers, read not only by radicals such as myself but by average white Americans trying to understand what was going on. Both Cleaver and Jackson joined the Black Panther Party, helping to increase the party's standing as "the vanguard of the black revolution."

Inside the prisons a state of war prevailed, with black-white gang battles, guards shooting selected activist prisoners, retaliation against guards. One of George Jackson's organizing partners in the Black Guerrilla Family, a self-proclaimed Marxist revolutionary group, was murdered by a guard who was then exonerated. A few days later, another guard was thrown off a tier balcony in apparent retaliation; Jackson and two other prisoners were charged with the killing.

In early August 1970, George Jackson's sixteen-year-old brother, Jonathan, invaded a courtroom in San Rafael, California, Marin County, where some black San Quentin inmates were on trial and held a shotgun to the head of the presiding judge, making him a hostage for George's freedom. In the ensuing gun battle, the judge, Jonathan, and two of the codefendants were shot. It was a bloody mess.

Out of the event, a number of people became fugitives, including Angela Davis, a young and articulate black Communist professor, who was charged with supplying the weapon to Jonathan Jackson. She became a notorious fugitive, the object of an enormous international support movement, who was eventually captured and acquitted.

In our tribe house on Pine Street, San Francisco, only a few miles from the Marin County Courthouse, we held a meeting right after the shooting. *What could we do to show solidarity with Jonathan Jackson and the black revolutionary prisoners?* The organization had developed some expertise in building bombs; several had been exploded successfully in New York City back in June 1970. Despite our dysfunction, our collective of the Weather Underground decided we would demonstrate our solidarity by blowing up the Marin County Courthouse.

I was given the job of casing the site, trying to figure out where and how to plant the bomb. This was the first overtly political act I'd been engaged in in months. I drove out to Marin in an old '57 Chevy station wagon, an anonymous car at the time, and parked in the courthouse parking lot. The building, designed by famed architect Frank Lloyd Wright, was a treasure in concrete of open California design, with colonnades and porticoes along its south-facing side to protect against the glaring August sun. Security, I found, was remarkably lax, given the events of the week before. No one stopped or even noticed me.

I ducked into a men's room and looked around. In a stall I found a drain that could be accessed by a screwdriver. I pried it up and peered down. It was two inches in diameter and appeared to go straight down at least a foot. I had the spot for the bomb. It was so simple.

I reported back to the collective. From there, other people took over the manufacturing of the device. A few days later, the bomb was put in the bathroom, with a timer set for after the building was closed. We called in a warning late that night, giving them time to evacuate the building. Then we waited to hear a report of the blast on the radio: nothing. The next morning one of us cruised by the courthouse to see if there was any evident damage: nothing. Worried that it would go off accidentally and harm an innocent visitor or a janitor, we called in a second warning, telling the security office at the jail that there was a bomb in the men's-room drain. It never exploded.

George Jackson was murdered by guards at San Quentin prison a week later, in an alleged escape attempt, and became another revolutionary martyr. He was buried with full Black Panther honors.

The Weather Underground branch in San Francisco was particularly busy that August: A plan was being hatched to help Timothy Leary, the pioneer and guru of acid, break out of federal prison at San Luis Obispo, down the coast. Leary was serving a ten-year sentence for possession of two marijuana roaches that agents had planted on him. Leary's religious and business associates—manufacturers and distributors of LSD, known as the Brotherhood of Eternal Love—put up the twenty-five thousand dollars for expenses (we scored a decent profit on the deal).

The leadership had made the decision to take on the action; I was never consulted. But I did help out in the physical preparations, putting together an escape vehicle and a safe apartment. Breaking Leary out of prison was a perfect accompaniment to the song Jeff and Bernardine had been singing since May, about Weatherman being a part of youth culture, not outside it. To them, youth culture—being about drugs, music, having a good time—was inherently revolutionary *and* political.

I tended to see it differently: We were opportunistically glomming on to the counterculture. For years SDS had stood to the side, criticizing the hippies for not being political enough. The Leary jailbreak appeared to me to be a transparent attempt to insinuate ourselves with our potential base, the flower children. Still, I put up no opposition to the Leary action, feeling neither clear enough to criticize the organization's opportunism nor strong enough to present an alternative view.

With six hundred dollars of the Brotherhood's money, we purchased a used three-quarter-ton 1964 Dodge pickup with a camper on the back. I was assigned to work with C. Van Lydegraf, the oldest guy in the organization, to get the truck ready. Van would be driving it to pick up Leary outside the prison and then take him north to San Francisco to an apartment I was preparing in the Mission District.

Van was a lifelong Communist in his late fifties, a union organizer from Washington State. As a young man, he had worked on the building of the Bonneville Dam on the Columbia River, and he had served for years on the Central Committee of the Washington State Communist Party. He had courageously faced down the House Un-American Activities Committee during the McCarthy era of the early 1950s. When the Sino-Soviet split occurred in the early 1960s, he sided with the Chinese against the Russian "revisionists" who wanted peaceful coexistence with the capitalist West; he left the party and helped form the Maoist Progressive Labor Party with other defectors. Later, Van left PL when his militant anti-imperialist position was shunted aside. In 1966 he hooked up with the Seattle SDS people as their resident guru, which is where I first encountered him. He had written an influential pamphlet that Weatherman circulated, called *The Object Is to Win,* the point of which being the need for the New Left to develop the capacity for armed struggle to overthrow the state. Unlike most old Commies, Van was still hard-core; he had not gone reformist or soft.

Van didn't care for the Leary action at all, instinctively not trusting anything that wasn't overtly and explicitly political, but he never said a word to me about it, because he was used to following party discipline. My impression was that he had some sway with Jeff, Bernardine, and Billy, certainly much more than I, and had said his piece to them. Then, as a loyal Bolshevik, he accepted the leadership's decision and shut up about his criticisms, in order to not undermine party unity.

Van and I got to work on the Dodge, which needed a new rear end. We jacked up the truck right out on a street running by the Panhandle of Golden Gate Park and swapped out the rear axle. I had never done any work on a car before, but Van had been repairing old trucks all his life. I was his helper. I liked listening to the old guy; it calmed me down and made me think maybe I wasn't totally off my rocker. It's inspiring to hang out with a steadfast Communist who'd never given up, a factory worker, a union organizer fired time and again. As a party official, he was committed to the core to follow the twists and turns to which history forced Communists to adapt. Van's participation in the Weather Underground was the culmination of his revolutionary dreams. Plus, occasionally, a young woman would sleep with him, in the spirit of comradely nondiscrimination.

Working under the truck, we constantly used tools from his green metal toolbox, which was about the size of a small tackle box. He had every cubic inch packed with hand tools and nuts and bolts. Van noticed me examining the toolbox and said, "If you want to be a revolutionary, learn a trade. I supported myself and my family as an appliance repairman, since I couldn't get any other job. There was a good fifteen years of my life when the FBI had me fired before I was even hired. That toolbox is how I earned my living."

I wondered where his wife and kids were. "Back in Washington State. They got the house. They don't see things quite like I do," he said sadly.

The other old guy in the operation, Tim Leary, was fifty. At that moment he was in prison several hundred miles south, training himself physically and mentally to scale the minimum-security facility's wall, then go hand over hand on a hot electrical power cable for about two hundred feet, then drop down fourteen feet or so. One moonless night he actually did it, as planned, and then, with an ankle sprained from the fall, ran a distance of about a quarter mile out to the highway. He crouched down behind a tree until a Dodge pickup with a camper on its bed happened by and signaled with its lights. Leary got into the camper, where a specially trained crew of two laughing hippies introduced themselves and immediately gave him a makeover—dyeing his hair black, outfitting him in new clothes for a fishing trip, and supplying him a brand-new identity, with papers to match.

A few miles down the road, the helpers got out of the camper and into a second car headed back in the direction of the prison, taking Leary's prison clothes to where they would be found in a gas station restroom garbage can 150 miles south of San Luis Obispo. Leary got into the front seat, and the two phony fishermen made their way north, canoe lashed to the top of the camper. The two old veterans of different wars spent the next hours as they slowly approached San Francisco vehemently arguing about the deficiencies of the youth culture. Van couldn't help himself, and he unloaded all his criticisms on the poor, unsuspecting acid guru who'd just escaped from prison. I wish I'd been there to observe that battle of the titans.

Van brought Leary to the apartment that I'd fixed up as the first safe house of the escape. The Weather Underground leadership all partied down with Leary that night, but I didn't even want him to know I'd worked on the action. For me, meeting him was out of the question. We were supposed to operate on a need-to-know basis, and Leary didn't need to know I was anywhere around.

The leadership worked with him on writing a joint statement, which the organization immediately distributed via the alternative press. "Brothers and sisters," Leary proclaimed to the world, "at this time let us have no more talk of peace. . . . If you fail to see that we are the victims—defendants of genocidal war—you will not understand the rage of the blacks, the fierceness of the browns, the holy fanaticism of the Palestinians, the righteous mania of the Weathermen, and the pervasive resentment of the young."

Leary's bombast escalated. "You cannot talk peace and love to a humanoid robot. . . . To shoot a genocidal robot policeman in the defense of life is a sacred act.

"Listen Americans. Your government is an instrument of total lethal evil." I noted what a nice line that was.

For our part, Bernardine added to the communiqué that Leary was a political prisoner for his revolutionary drug views, and, naming black revolutionary prisoners like H. Rap Brown and Angela Davis, she pledged the organization to "the task of freeing these prisoners of war." I don't recall any other prisoners ever freed by the Weather Underground, especially any black revolutionaries; they didn't tend to be held in minimum-security lockups.

"We are outlaws; we are free!" crowed Bernardine's part of the communiqué at the end.

Within a few days, Leary arrived in Algeria for a momentous unity celebration signifying the marriage of the youth counterculture with the political struggle for national liberation. Appearing with him in a joint press conference facilitated by Weather friends and relatives from the United States was the exiled star of the black liberation movement, Eldridge Cleaver, minister of information of the Black Panther Party and a fugitive wanted for the wounding of an Oakland policeman. Cleaver was not only one of the three top Panther leaders, along with Huey Newton and Bobby Seale, but also a bestselling writer, author of the prison memoir *Soul on Ice.*

The Leary and Cleaver marriage didn't last long. The two didn't get along at all: Cleaver became embarrassed when he finally figured out who Leary was. (After confiscating Leary's drugs, he wouldn't share with him, either.) Right about the same time, the Algerian government caught on to Leary's being the apostle of acid liberation, and no matter how revolutionary and anti-imperialist they were—having kicked France out of their country just a few years before—they didn't want to be associated with LSD or any other drug. So the Algerians and Cleaver booted Leary and his wife from Algeria. Leary's next stop was a press conference in Cairo, in support of the Palestine Liberation Organization, accompanied by our Weather friends.

Cleaver at that time was in the midst of a split with the Oakland, California, headquarters of the Black Panther Party over the issue of armed struggle. Cleaver favored "picking up guns" against the government of the United States, but by now Huey Newton, the founder of the Panthers, had come to the more sensible conclusion that it was a losing proposition and that service programs like breakfast for children, not shoot-outs, would build the Panther base. Cleaver must have initially thought that the alliance with the jailbreaking Weathermen was somehow useful to him, as it bolstered his argument that the revolution was happening. (Amazing, when you think about it, the Weathermen being used to support one faction of the Panthers against another.) Cleaver's closest followers and allies in the BPP were the New York City Panthers, many of whom were still under indictment in 1970 in the Panther 21 case. Later they would become the basis for the Black Liberation Army.

Leary and his poor wife, Rosemary, wandered around for several years, stateless and homeless, finally landing in Afghanistan, where the hash was excellent. In 1973 he was apprehended there by U.S. agents who hauled him back to the United States. He then told the agents of the "instrument of total lethal evil" everything about the escape, even naming as many names as he could remember, including mine. (Had my comrades bragged to him that Mark Rudd worked on his escape?)

No prosecutions ever resulted from Leary's snitching, however. He was the only witness against us, so the federal attorneys probably figured they couldn't get a conviction. Since Leary claimed he'd taken thousands of acid trips, the feds could hardly turn around and offer him to a court as a credible witness. For some reason, probably because I like audacious old charlatans, I've never felt angry or vindictive toward him. I exempt Leary from the principle "Thou shall not rat out thy comrades." He died in 1996, an apostle of space colonization, having transformed his identity multiple times in the previous two decades. Seven grams of his ashes were rocketed into space. Eldridge Cleaver died in 1998, a Republican and a Mormon at the time, after having tried various brands of evangelical Christianity.

Despite the excitement over Tim Leary's escape, I was sinking deeper and deeper into depression in the fall of 1970. Along with its delights, San Francisco also has its noir side that can chill the soul. Every morning at 5:00 A.M., the temperature clock at my bus stop on Geary Street would read a frigid fifty-three degrees in the fog, as I made my way down to the union hall looking for work. For the first time in my life, I thought about suicide. This is something Jewish-American princes generally do not do.

I ascribed my problems to my own failings: I wasn't brave enough or committed enough to the revolution, so I couldn't adjust to life as a fugitive. I was ashamed, too, of my emotional response to my fall from leadership, ashamed of my hurt and anger and feelings of victimhood. I was unsure of who I was or how I should act.

Finally, on New Year's Eve, Sue and I were holed up in a tiny $7.95-a-night hotel room in the Tenderloin. I had brought along a small table radio, a bottle of wine, two hits of acid, and some grass. In the middle of our acid trip, Sue asked me, "What are you doing? It's like you're on a treadmill. You're just stuck, not going anywhere. You need to change your life, get away from the organization. There's no air in the Bell Jar."

I instantly realized she was right.

"Why don't you separate from the organization, go out on your own? You can evaluate things from a new perspective," she suggested.

This made sense, but I was scared. How could I survive, a fugitive alone? I needed the resources and support of the organization, I thought, at least for security, such as acquiring good ID. Like a prisoner contemplating his release from prison, I wondered if I could make it on the outside. Sue and I talked it out all night, as I went back and forth in my mind.

By New Year's Day morning, after the acid had worn off, I resolved to take the leap and leave the organization. The thought was terrifying but cleansing at the same time—a new beginning might save me.

That was, incidentally, the last LSD trip I ever took.

14

Santa Fe

On a bitter-cold day in late January 1971, I was waiting outside Bookbinder's, a restaurant in downtown Philadelphia, my safe city from the year before, hoping to God that my parents weren't tailed. *Please let them follow my instructions to ditch their car and take a taxi.* At last a middle-aged couple, all bundled up, walked toward me down the sidewalk carrying a paper shopping bag. Jake, my father, was burly, not quite six feet, wavy black hair cut short beneath his Russian-style fur-and-leather hat; Bertha, almost a foot shorter, stocky, was wearing a heavy coat with a fur collar and sensible fur-lined boots. They recognized me finally, beneath my new long hair and beard, and we embraced. It'd been May the year before since we'd seen each other. Jake was stooped over more than I remembered him; he seemed older and smaller.

We went inside the huge restaurant for a late lunch and allowed ourselves to be led to a white-clothed table. We took off our coats and hats and stared at each other, my mother on the verge of tears and Jake just glum.

"It'll just make things worse to cry," I said, "especially since there are so many people here."

"I cry every night," she said, "and in the morning, too."

"For the last two years, things have been getting worse and worse," my father said. "It was bad enough when you were kicked out of Columbia, but then the Weathermen was even worse. Now, being a fugitive . . ." He trailed off, unable to finish.

I was at a loss as to what to say that I hadn't said before.

My mother took up her old thread. "We have the radio on all the time, just hoping we won't hear that you've been shot by the police. I don't know, maybe it would be a good thing if you were arrested. Or turned yourself in."

"I'd die in prison," I said, "so get that out of your heads. Besides, I'm not going to give the government a victory." I wanted to change the subject and said, "I have some good news. I'm leaving the organization, getting away from my old friends. I'll be on my own, but at least I won't be making any bombs."

"What will you do?" my father asked.

"Out west I can lose myself all sorts of places. I haven't figured it out yet, but my plan is to get a vehicle and travel around until I find a place where I'll be safe. I'll work and be quiet for a while, try to think things out."

"We brought you three thousand," my father said.

"You didn't withdraw the money from your account, did you?" I asked.

"No, I accumulated it in cash over the last year, from the friends in South Jersey."

I knew what he was referring to. Years ago, when my father had worked for the Army Exchange Service, he had helped a friend get the base PX jukebox concession. Ever since, he had received a small cash kickback every few months. It was the only illegal or unethical activity I'd ever heard Jake involved in. In our family the small rake-off wasn't even considered a crime, just a normal mode of doing business, the single atavism from an earlier, poorer time. It came in handy now, though, to have a hidden and untraceable source of cash.

My mother, as always, said what was on her mind. "I don't know if we should be giving you this money. I just don't want you to have to take more risks, like robbing a bank or dealing drugs. If this keeps you from getting caught or killed, I'm happy."

In my family there have always been three things that are interchangeable—food, money, and love. Visitors to our house were never asked, "Are you hungry?" which Jake and Bertha would have thought was impolite. Instead the visitor was sat down at the table, no exceptions. Both of my parents would then compete to pull food out of the refrigerator at random—leftover roast, chicken soup, cheese and crackers, whatever there was.

They were equally generous with money, never denying me or my brother, or other members of the family. Our family practiced a primitive form of Communism: If anyone in the extended family needed the basics—a car, a house, education—they'd get it from the most financially successful ones, Jake and Bertha. No one went without. Despite achieving financial security, my parents still lived in the same little ranch house and still drove the same Fords. We weren't even close to the league of the flashy nouveau riche Jews who we saw driving by our house in their Cadillacs, on their way up the hill to South Orange. But growing up, I had wanted for nothing. When I was into ham radio, at the age of eleven, my father bought me the gear, telling me that the only thing he asked in return was that I do the same for my kids.

But here I was, at age twenty-three, still their kid, desperately needing their money. Taking from them was the safest and quickest way for me to detach myself from the organization. For months in San Francisco, I'd been broke, not even accumulating enough to buy a car by working or getting handouts from aboveground "contacts." In fact, I had borrowed money from the organization for airfare to get back to Philadelphia for this meeting.

Money was also a way of keeping my connection with my parents. As self-serving as it may sound, taking the money was akin to accepting their love. I could no more reject their money than I could their love.

"We sold our house," my father said. "We weren't safe anymore, exposed out on the corner of two main streets like that. Some nut who doesn't like you could throw a rock through the picture window. I'm building another one up on Ridgewood Road, on a little cul-de-sac called Washington Park. I designed it myself, and my men from Elizabethtown Gardens are building it. I'm the contractor, out there every day working with the crew."

I didn't know whether this was good or bad. He seemed to be blaming me, but at the same time he was excited and proud to be building a new house at the age of sixty. Maybe it distracted him from his worries. "That's great," I said. "When this is all over, I'll come and see it."

"When this is all over," my mother said, "we'll pack a basket and go off to the seashore." It was her favorite expression, from the movie *Never on Sunday*. Melina Mercouri's fairy tales always ended happily, the characters going off to the seashore rather than dying their normal grisly fairy-tale fate.

The restaurant was by now filling up with customers. We looked around. Suddenly the three of us realized that every table in the room was jammed with short-haired, blue-suited white men, government agent types. *Holy shit!* We appeared to be attending a convention of FBI agents or cops. My father called for the bill, paid, and we quickly left.

Out on the street, we said good-bye once again. This time something seemed very final about our parting. "Be careful," my mother said, tears in her eyes.

"I will, believe me," I replied. "I'm not as crazy as I was."

"Good. We always believed in you," my mother said. "I guess everything happens for the best."

"Everything happens," my father said, shaking his head.

Back in San Francisco, I found a truck on a used-car lot in the Mission district—a beautiful brown 1954 Chevy pickup, the first one made with a one-piece windshield. I paid three hundred dollars of my parents' cash and drove off in the first vehicle I'd ever owned. After a week fixing up the truck, Sue and I set out for Death Valley, California, on a winter camping trip, our shakedown cruise. Unfortunately, the engine started belching smoke by the end of that trip, and after trying to do an overhaul, I wound up trading that truck for a 1952 GMC panel truck painted washed-out pink. The hippie I bought it from told me the truck's name was Scarlett.

With less than a thousand dollars left of the three thousand from my parents—I had given half to the organization as a good-faith sign of our connectedness—I took off for Oregon to find a safe place to settle, thinking that the rural counterculture there might prove congenial.

I learned from a hitchhiker about an area on the Rogue River in southern Oregon where there were several communes. What I found there was a collection of homemade dwellings—shacks and tepees—in various states of repair. I struck up a conversation with a young guy about seventeen or eighteen years old, a refugee from L.A. While we were chatting about how the commune was organized—loosely, with a communal house for those who wanted to spend time together—I noticed that his whole body was covered with open and weeping sores. I asked him if perhaps they might be staph infections. He replied that he was on an all-raw-vegetable cleansing diet. "That's just the poisons coming out of my system," he told me. I got back on the road.

The hippies I encountered in rural Oregon were too spaced out for my taste. This was 1971, the war continued to rage, and I couldn't feel sympathetic with those who had chosen escape. I was still a partisan in the old cultural-versus-political split, despite the new Weather-hippie line. Visiting Eugene and Portland, I found good people, urban activists who were more involved in building community with political edges such as food co-ops and newspapers. But I had to stay away from these vestiges of the movement out of fear of being recognized. Nothing felt right for me, alone in the rain.

I called Sue back in Oakland and asked her if she'd join me, so we could live together. In joining me underground, Sue would have to change her identity, take on false papers, leave her family and friends behind. We assumed—rightly, as it turned out—that the FBI sooner or later would get around to checking up on her, following leads that might lead to me. They had already visited my relatives and several other friends. Very soon they would get to her folks, conservative Missouri Republicans whom we were afraid to trust with our secret.

Sue was willing, had been for months. I'd been the archetypal reluctant male, unwilling to make the commitment. However, she had one condition: We would have a child. I thought the idea was completely absurd, given that we were fugitives, with no security in our lives. Only a year before, I had believed the revolution so imminent, and the struggle so intense, that there was no space or time for children in my or any other revolutionary guerrilla's life. In fact, at the craziest period of Weatherman, around the time of Flint, we had demanded of those who would join us that they give up their children for the sake of the revolution. Two or three people actually did it, very briefly, until they came to their senses.

Now, with Sue's guidance, certain older and saner feelings were filtering back to me: thoughts about children, and living in a house or an apartment, having a partner, longings that I can describe only as "normal." I was following my parents' patterns.

"Okay," I told her, "we'll do the baby."

Our life was not going to be conventional, though, since we still had both the terrible anxiety and the complicated problems of life as fugitives to contend with: how to get good ID papers, birth certificates, drivers' licenses, real Social Security cards; how to find a place to live; how to make a living. No matter what, things would be easier now, I thought, because I had a partner.

Early that summer of 1971, Sue and I traveled to Oregon, living in our panel truck, Scarlett, looking for that home we hoped to find. In Portland we heard from an organization contact that an extremely dangerous situation had developed back in the Bay Area. The apartment on Pine Street, where I'd lived for more than six months, had been raided. My friends who were living there had split in a panic just hours before the FBI got to the apartment, leaving behind a lot of important stuff.

Bernardine and Jeff and Billy Ayers had been followed from a Western Union office where they had picked up money sent by friends in the East. Though they escaped the tail, their car's license plate was noted, and that led in some way—perhaps via a parking or traffic ticket—to other people's IDs and eventually to the house on Pine Street. Too many items were linked together, and within the space of about twelve hours a whole painstakingly built network was rendered unusable.

No one was busted, but the event came to be known in the organization as the Encirclement. All the San Francisco organization's vehicles and apartments had to be abandoned. Sue and I were thankful we were already separated from the organization, but we realized that our IDs—actual California drivers' licenses with our pictures on them—might be compromised. We needed new identities, and quickly.

We were in the habit of using real licenses and real Social Security cards so we could drive and work. These could be built on phony birth certificates or birth certificates from babies who had died very young. Someone we met in Oregon showed us a New Mexico driver's license with no picture on it, unlike the one from California.

Licenses without pictures were golden, since pictures in Motor Vehicle Bureau records could at some future time be used to find us. Though I had changed my appearance with a black Afro, my hair dyed and permed into tight, tiny curls, I had no confidence in the disguise, knowing that it didn't fit my overall coloring or facial and body hair. Also, the perming and dyeing process was excruciating, with stomach-turning chemical odors I gag on even to this day. And I had to redo my hair once a month. I longed for the time when I could live without that horrific disguise.

None of the other western states had licenses without pictures, so we happily set off in Scarlett for New Mexico, glad to be done with the rain and Oregon's general white-bread quality. Slowly making our way through Idaho, Utah, Wyoming, and Colorado, we nursed the old truck with its leaky radiator. At one point we had to stop and do a valve job, a major undertaking I accomplished with advice from helpful mechanics. We camped along the way, worked at odd jobs such as fruit picking, and made quick runs through Boise, Salt Lake, Cheyenne, and Denver.

After several weeks we crossed the Raton Pass from Colorado into New Mexico. Within hours it hit us that we were home. It was in part the smell of the desert—the fragrance of sweet sage, the summer-afternoon rains, the clean mountain air lit by a golden light. I loved the desert aridity: homes made of adobe, sun-baked mud; the preciousness of the sparse water in the rivers, streams, and irrigation ditches.

Taos was our first stop. Beautiful as this little town was, nestled in a mountain valley with its high-altitude light, it seemed wrong for us—too small, too easy for me to be recognized. Also, it didn't take us long to learn that there had been a history of conflict—almost a war—between the indigenous Chicanos and the hippie immigrants. People resented the sons and daughters of the white *gabacho* (honky) middle class playing at being poor while living on county food stamps and welfare. To many Chicanos who had lost much of their land and suffered more than a century of racism, this was merely the latest gringo invasion. They fought back. While sitting naked in an idyllic hot spring in a cottonwood canyon outside Taos, we were calmly told by other bathers that several hippies had been shot at that spring recently. We got back in the truck and headed south.

Just sixty miles down the highway was Santa Fe, the state capital, whose population at that time was a mere thirty thousand. An art and tourist center, it is one of the oldest cities in the United States and, in a word, gorgeous. Despite its small size, we thought, optimistically, that we could live here and not be noticed.

Sue and I checked out the Santa Fe "motor buricle" office, as we called it, and found, to our disappointment, that the state had just switched to Polaroid photos on their licenses, copies of which we assumed were kept in files somewhere. We decided to go ahead and get the licenses, even with the pictures, since living in New Mexico seemed worth the risk.

We found an apartment in an old adobe house that had been converted into a duplex. Located in the West Side barrio, the three rooms rented for a hundred dollars a month, utilities included. Much to our delight, our landlord, Don Salomon Mendoza, an adobe builder, hired me right away, working for the minimum wage, $1.90 per hour. With a twinkle in his eye, Señor Mendoza would refer to me as his "hippie *peón.*" *Peón* in Spanish means about the same thing as it does in English: a menial laborer.

Señor Mendoza was in his late sixties, small, wiry, naturally speedy, hyped up on his work like so many other builders I've known since. As a young man, he had come to New Mexico from Mexico to work in the coal mines south of Santa Fe. He could very easily tell I had never worked in construction before. A real Laurel or Hardy character, I'd carry five-gallon buckets of water and spill half along the way. Once he sent me to the building-supply house in his flatbed truck to pick up a roll of stucco wire. By the time I got back to the job sitc, thc roll had, of course, fallen off.

One day he stood watching me sweeping up the house we were working on. I was making more of a mess than I was cleaning. He took his Robert Burns cigarillo out of his mouth and said to me, deadpan, "You know, it depresses me to watch you." We both laughed.

Señor Mendoza and his two sons, Jackie and Charley, taught me how to mix concrete, put up drywall, do elementary carpentry and masonry. I was a carpenter and mason's helper, doing the jobs local Chicano men learned when they were kids, only I was twenty-four years old. When the old man was elsewhere, his sons and I would smoke weed together as we plastered, giggling and singing Marty Robbins's "El Paso" in our best cowboy-pachuco-hippie tones.

The Mendozas had Model Cities contracts to rehabilitate houses in the barrio where we lived. I would often find myself in elderly Chicanos' homes, getting to know them as people and as friends, beginning to appreciate the Chicano language and culture.

I was Tony Goodman in Santa Fe, and Sue was Flan Taraval, the last name derived from the trolley line in San Francisco. I chose "Goodman" for obvious reasons and because it sounded Jewish, and "Tony" because I liked the sound, it was easy for people to remember, and because it was vaguely *not* Jewish. I was Mr. Ambiguity. No one ever asked us where we came from, or why, or directly questioned the vague cover story we offered: We're from San Francisco, tired of city life, exploring the West. At first, surprised about this lack of curiosity, I chalked it up to the unique tolerance of Santa Fe, which had seen generations of artists and other outcasts from straight America. But after living as a fugitive for years in other parts of the country, I realized that there is a national characteristic of not inquiring about people's pasts. Perhaps it's due to politeness, or possibly just lack of interest.

Sue and I fit in well with the hippie community, of course. We got involved in the initial stages of organizing a food buying co-op called Fat Amerika. We made close friends in our neighborhood, partied with them, and felt at home. We had two dogs and a cat named Jones—after Jeff Jones—all substitutes for the baby that was not happening. We joined a karate club and enjoyed the small cross section of Santa Fe society we met there.

We often talked politics, swapping experiences and observations with our new friends, some of whom were refugees from radical scenes at college campuses on the coasts. We didn't hide our politics, because we believed that the best strategy for survival was to tell as much of the truth as possible, which was Raskolnikov's philosophy in Dostoyevsky's *Crime and Punishment*. I conveniently ignored the fact that Raskolnikov got caught—he turned himself in.

Occasionally my truth policy did get me into trouble. When Nixon and Kissinger bombed the harbors and agricultural dikes of North Vietnam in April 1972, a big escalation of the air war, demonstrations spontaneously broke out around the country, including at the University of New Mexico in Albuquerque, where students blocked traffic on a nearby freeway. It was similar to the reaction to the invasion of Cambodia in 1970. The disorganized radicals of Santa Fe even held a rally in the plaza at the center of town.

Several of my new friends tried to enlist me in plans for the rally, and also to go to Albuquerque to join the students there. I had to make excuses. "Got to work," I mumbled, or "Can't get away," really lame stuff, especially for a person who talked such a good game about imperialism and revolution. I hid out during the demonstrations, which caused me to question my purpose in being underground. *What, exactly, was I accomplishing in my life as a fugitive, other than not getting caught?*

From time to time, people from "the Eggplant" would stop by our apartment in Santa Fe to check up on us. (Sue had coined that code word for the Weather Underground from the obscure novelty song "The Eggplant That Ate Chicago" by Dr. West's Medicine Show & Junk Band.) An old friend from San Francisco brought news that the organization had dispersed in the wake of the Encirclement. Most people were now living on their own, working, integrated into countercultural communities, as we were. The political work of the organization—communiqués, development of aboveground contacts, bombings—had almost ceased.

But not quite. In September 1971, inmates at Attica Correctional Facility in New York State took over the prison. Governor Nelson Rockefeller sent in state police troops, and they killed fifty-three people, including some of the inmates' guard hostages. Sam Melville, a white revolutionary many of us had known at Columbia, was among those murdered by the police. A convicted bomber, he had been one of the few whites on the Attica prisoners' strike committee. After the state police regained control of the prison, they sought him out and shot him point-blank, along with several other leaders.

In retaliation the Weather Underground bombed the offices of the New York State Corrections Commission. No one was injured; it was just property damage. At the time I felt that seemed an appropriate revenge for the Attica massacre.

One day in April 1972, I read in the local Santa Fe newspaper that Dr. J. S. Horn, an Englishman who had worked for some years as a medical doctor in China, was going to speak at a downtown church. Sue and I had read his book, *Away with All Pests,* about practicing medicine in revolutionary China. We wanted to hear him speak. We generally stayed away from any movement-type events, but this one seemed safe enough, as it was sponsored by a local community clinic and was not explicitly radical. We invited our neighbors to go along, a couple with whom we'd become close, though they didn't know our true identities. The church was full, perhaps two hundred people were there, but, glancing behind us, I immediately saw a small woman with short brown hair. I couldn't help but stare, more out of amazement than anything else: It was Jane Alpert, now a fugitive due to the bombings she, Sam Melville, and others had done in New York, back in 1969. She and I had had a brief fling in early 1970—a one-night stand complete with acid trip—and I had helped talk her into jumping bail and going underground before her sentencing.

What was most disturbing about seeing Jane, and probably the source of my astonishment, was that she had absolutely no disguise. Her hair, her clothes, her mannerisms were precisely the same as in New York City in 1969. Along with changing my name several times, I had already gone through three disguises in two years. In Santa Fe I had very long blond hair, dyed with L'Oréal #9A Light Ash Blonde. I kept it up religiously, whenever my brown roots started showing. It was something I hated but considered necessary.

I had heard reports about Jane from our occasional Eggplant visitors. People were always sorted out into two vast categories: Great and Not Doing Well. "Great" meant that the individual was a loyal supporter of the underground and in total agreement with whatever line the leadership was pushing at the time. "Not Doing Well" could mean anything from political differences to losing a job to being suicidal. (I suspect I had been permanently placed into the NDW category quite a while before.) The reports on Alpert were always Not Doing Well. My first glance confirmed this: No disguise meant that she was lonely, unhappy, and probably unconsciously wanted to get caught.

During the talk I kept staring at Alpert, first checking to see if it was really she, then trying to get her attention. Toward the end she got spooked and fled the room. As soon as the event was over, I went up to the woman with whom Alpert had been sitting and told her I was an old friend of Foxy's (the name I'd known her by) and that my name was Tony and I'd like to talk with her. She said she'd pass along the message.

We met later that night at the deserted plaza. I was cautious, wondering what state I'd find her in. I began by telling Jane that I liked her introduction to Sam's letters, referring to the essay that had just been published in a book of Sam Melville's letters from Attica.

"I'm sorry I wrote it," she said. "If I had it to do over again, I'd come down on his sexism." I was surprised. The essay had praised Sam as a committed revolutionary who had remade himself in order to fight against the war, racism, and imperialism. It was a loving piccc, one that revealed how deep was Jane's loss.

"I've been getting in touch with my own oppression as a woman, and I don't have much sympathy anymore for an oppressor like Melville," she continued. A light of recognition went on in my head; that explained her anger toward Melville: Jane was into the feminist line of opposing the sexism of men on the Left.

We talked for over an hour. Alpert described the difficult time she'd been having in the two years since she had jumped bail: loneliness, isolation, constant moving, loss of friendships, guilt over her parents' suffering. I told her I had taken a leave from the organization and was living with a woman here in Santa Fe, was fairly well settled, working, with new friends. She seemed envious.

We made a date to see each other again. Before parting, she told me that after I had left the lecture at the church, a young man had come up to her friend and inquired, "Who is that guy? I think I know him from Columbia SDS."

Great. This was the third time I knew of in less than nine months in Santa Fe that I'd been recognized. The other two hadn't turned out badly, though they did cause us some anxiety: They were people I had known from high school and college who I felt wouldn't betray me. But three times was just too many.

Sue and I stayed up all night trying to sort out our problems and our options. We huddled together in bed in the dark. We had to whisper so our neighbors on the other side of the thin wall wouldn't hear us. Sleep was impossible anyway: We were both sick to our stomachs with fear.

"Shit, what a mistake going to that stupid lecture was. Now we're fucked," I said.

"We'll just have to leave Santa Fe," said Sue. "There's no way around it. If this person who recognized you is dangerous, or just untrustworthy, we'll need to be gone tomorrow. And ditch the car and our IDs."

This prospect put us into a miserable funk. We had so much going here and had worked so hard to put it together. I was even due to start a new job that paid better, with a different contractor, in two days.

"Maybe this guy who recognized me won't be so bad. I'll find out who he is first thing in the morning and go by his house to check him out," I said.

Sue thought about that. "It's risky," she said, "but so's packing up and leaving in a day. If we can take our time, it'll be safer."

"We can tell people we're going back to California. Somebody there can develop a sudden illness, like a terminal case of leukemia," I said.

"Do we have to make the story so depressing?" Sue asked.

"The whole thing's depressing," I replied.

"What about Jane Alpert? What do you think is going on with her?"

"She appears to be alone," I said. "She's way down. She told me she had a clandestine meeting with her father here in Santa Fe a few weeks ago. That's really risky."

Sue had never met Foxy.

"Maybe we should befriend her, get to know her better. Maybe we can help her and keep her going for a while," I proposed.

"Now, that's risky, no disguise, depressed, denouncing Sam Melville," Sue said. "I wonder if we'll be able to do anything for her. Maybe she wants to get caught."

"You and I have each other," I said, holding her.

In the morning I made some inquiries and found out the name and address of the guy who had recognized me. I remembered the name from Columbia: He'd been a freshman in September 1968, after the strike, and had joined the SDS chapter immediately. From what I remembered of him almost four years before, he was okay, but something of an ego freak, and I was scared he couldn't be trusted to not brag around that he'd seen Mark Rudd in Santa Fe.

I went over to the guy's house and introduced myself as Tony Goodman. I told him I had worked in the New York SDS Regional Office. He showed no hint of knowing who I really was, seeming to accept the story I was concocting on the spot. I suggested that we might have met each other at a Panther 21 defense rally.

I wondered while I was spinning this whole story whether it would be better security just to tell him who I was, then pledge him to secrecy. I chose to stay with the story, convinced somehow that he really didn't know my identity. It was a tough call— maybe he was fooling me.

I reported back to Sue, and we decided we didn't have to flee immediately. But we agreed that Santa Fe was just too small and too much a mecca for counterculture and radical immigrants. We loved the place and didn't want to leave, but there was no future here.

Within a couple of weeks, we'd made a quick trip by car to Los Angeles, where the California organization—and the leadership—was regrouping. We wanted to talk with them about our predicament. Our idea was to go east. California was just too hot. As a consequence of the Encirclement the year before, the feds had convened an investigative grand jury in San Francisco, which had called seventeen witnesses, most of whom had no connection to the Weatherman. The grand jury turned out to be a great show of solidarity with the underground, since no one cooperated. Still, we knew that the hunt for us was on, and we needed to be careful.

We had old friends in the New York area with whom we wanted to stay in contact. We also thought it would be good to move closer to the Eggplant. New York and Boston were the only centers left outside California, the organization having by now shrunk to a tiny fraction of its former size.

In L.A. we found that the remaining weather fugitives were living in collectives that were more like shared apartments than political units. People's energy was consumed by the tasks of daily living as fugitives.

When we described our situation in Santa Fe, Bernardine and the others readily agreed to help us relocate to the East Coast. We would move that summer.

Back in Santa Fe, Sue and I befriended Jane Alpert, hung out with her at our house, and took day trips with her around the area. We learned more about her life, told her about ours, including that Sue was trying to conceive. At one point, for reasons I can't begin to remember, I found an old rifle at a flea market, bought it for twenty-five dollars, and gave it to her.

I worked that whole spring for two Anglo guys not much older than I who were building an adobe house on speculation. The crew was a fine representation of the male hippie community in Santa Fe—funny, laid back, somewhat politically aware and radical. There was even an ex–Progressive Labor guy on the crew, but we never got into a fight.

The bosses, on the other hand, had a thin patina of hipness and social concern that was betrayed by their deeper nervousness and money-grubbing. I watched those two contractors push us around, get verbally abusive, even fire people. Sometimes it had to do with covering up their own incompetence and lack of experience. As the weeks went by, the crew grew more and more disgruntled. There didn't seem to be a lot of difference between hip capitalism, as practiced in Santa Fe, and regular nasty capitalism.

I was planning to quit as of June 15, but a week earlier it suddenly occurred to me that I'd had enough. I told one of the bosses I was finished. As I packed up my tools, I just started laughing, hysterically, uncontrollably, and then continued even as I walked to the highway to hitchhike back to town. I had a grin on my face all day. At a party a few days later, I heard from friends on the crew that the bosses had both picked up hammers, thinking I'd flipped out and that they might need to defend themselves. And all I was doing was laughing.

Not long after, Sue and I packed up our gray 1966 Plymouth station wagon, bought surplus from the U.S. government for six hundred dollars, with our two dogs, Ernie and Petra, and Jones the cat, and set out for the East Coast, camping along the way. We vowed we'd come back to New Mexico someday. On top of the car was a huge box I had built out of plywood to carry the few possessions we were taking. Smuggled among the essentials was a stash of dried red chilies to help us survive the deserts of the East.

15

Schoolhouse Blues

Leaving New Mexico in the summer of 1972, Sue and I landed in an abandoned century-old redbrick one-room schoolhouse in western Bucks County, just south of Allentown and Bethlehem, Pennsylvania. We had chosen the area because it was within striking distance of both Philadelphia and New York City. The owner agreed to let us make it habitable, and in return we paid $150 per month rent. He provided the materials, and we hauled out mountains of trash, installed plumbing, electricity, a coal heating stove, new plaster, taking almost a year to remodel the interior. Learning as I went along, I built a loft and staircase for our bed, a kitchen, as well as a bathroom. The ceiling was fourteen feet high, with eight-foot-tall windows on three sides. We lived in an architectural gem.

Very quickly we made new friends in the area—back-to-the-land college dropouts like ourselves—with whom we partnered in a small business that manufactured wooden toys. We invested two hundred dollars for machines and became full partners in a woodshop that netted us about eighty cents per hour of sweatshop labor in a nasty old chicken coop. That Christmas was lean, because we had met all our holiday orders by the beginning of December and had no more cash coming in.

Money really didn't matter that much, since we were overjoyed that Sue was then three months pregnant. Our elation turned to misery on Christmas Eve, however, when Sue suffered a miscarriage. In the middle of the night, I raced her to the hospital in Bethlehem. In our minds we connected the loss of our baby to the Christmas bombing of North Vietnam.

Two months before, in October 1972, the United States and the Vietnamese had almost signed a peace agreement, but at the last minute Nixon balked. He and his secretary of state, Henry Kissinger, wanted to "send a message" to the enemy about their willingness to enforce the treaty. Like so many similar messages coming out of Washington over the years, this one was written in blood and fire. For twelve days in December, hundreds of B-52s flew more than three thousand bombing sorties against Hanoi and Haiphong, the two cities of North Vietnam. It was Nixon and Kissinger's ultimate Christmas card, death from the skies.

The *New York Times* called the bombing "terrorism on an unprecedented scale" and "Stone Age barbarism." The *Washington Post* added the adjectives "savage and senseless." That was exactly how the Christmas bombing—as it came to be known—struck Sue and me in our drafty little house. "Nixon killed our baby," we told each other. That cold New Year's of 1973 was a low point, with Sue convalescing in bed and me trying as best I could to minister to her.

A few weeks later, on the evening of January 27, 1973, the public radio station in Philadelphia, WHYY, broadcast seemingly endless ringing of bells, one after the other, to celebrate the signing of the Paris Peace Agreement earlier that day. Sue and I had made a special meal of steak and wine as our own little celebration. In a jubilant mood, we sat down at the table I'd made from an old cable spool, but when the bells started pealing, we suddenly found ourselves weeping in each other's arms, involuntary streams of tears for the millions killed and also for ourselves. Years of pain and hurt and frustration were loosed by the sound of those recorded bells. Their complex tones echoed for a long thirty minutes what had been the relentless beat of our lives: *Vietnam, Vietnam, Vietnam.*

We'd worked and waited years for this day, willing to do *anything* to end American involvement in this war—yet we'd never envisioned this moment. For all those long years, the end of the war was only a distant abstraction, never real, and even as the negotiations dragged on throughout 1971 and 1972, we still could not picture how it would end. Now U.S. combat troops were to be withdrawn from Vietnam, in return for which American prisoners of war in the North would be freed.

We cried that night of the signing of the peace agreement because so many had paid such a high price for this American crime, this unnecessary war. So many Vietnamese had died, so many Americans had had their lives maimed. Sue and I were war flotsam also, fugitives in a house in Pennsylvania, with nothing to show but our failed hopes for a child.

We also knew that the war was not over. Violating the spirit of the peace agreement, the United States had turned over all its bases to the South Vietnamese army. The puppet army had a million troops and the fourth-largest air force in the world but wouldn't have existed a day without U.S. funding and airpower. Despite some congressional noises about cutting off appropriations, the money to continue the war kept flowing.

So we cried that night with bittersweet emotions in the old schoolhouse in the wintry, bleak Pennsylvania countryside. It was the end of a war that was not at an end, a Vietnamese victory that was not yet a victory.

Around that time we realized that our toy business had no future. Sue got a job running a small gift shop in a nearby tourist town, and I answered a newspaper ad for a construction laborer. I lied about my experience but was hired as a mason's helper, at the then-huge sum of forty dollars a day, regardless of the number of hours worked. Even though my new boss instantly saw I knew much less about masonry than I'd claimed, he kept me on. He probably sensed I badly needed the job and was willing to learn and to work.

The boss, a guy in his late thirties nicknamed Sneezy for some unknown reason, was a subcontractor for a larger company that built in-ground concrete swimming pools. He and I, our entire crew, did exactly one portion of the job—setting the coping and tile around the pool edge. We would load Sneezy's flatbed truck with sand, cement, concrete copingstones, and tile at a quarter to six every morning, then head out for the job, a concrete pool shell that had been poured the day before in a residential backyard. We traveled daily within a 150-mile radius of Doylestown, Pennsylvania, which is about forty miles north of Philadelphia.

I picked up all the moves within a few days and was able to supply this hyperactive madman with his mud (concrete mortar) and stones (precast concrete blocks). Sneezy would be too wound up to stop for lunch, having worked himself into a frenzy all morning, so I learned to eat my sandwich at the mortar mixer between loads. Besides, it was to my advantage to get the job done quickly and get home by midafternoon.

Sneezy was a narrow, mean-spirited little bully who enjoyed tyrannizing those under him, such as his poor wife and kid—and me. If I was a bit too slow or the mix was a bit too wet, he'd bark at me, "Sucker, can't you do anything right?" as if this were the marines and he were a drill sergeant, which was actually what he *had* been until just a few years before. Several times a day, especially at first, I'd be told what a stupid numbnut I was (whatever that meant).

I'd just let it roll off my back. With no references and not much work background, I would have trouble finding a job this good, so I decided the first day to take whatever this lunatic dished out. In fact, taking Sneezy's abuse became a point of pride for me, my distorted Zen practice.

I even enjoyed knowing I was in a relationship (of sorts) with this right-wing jerk who had twice voted for Nixon and passionately hated the traitor peace creeps. I had long hair—compared to Sneezy's regulation marine flattop—and he had a very good idea what I was about. When we weren't choosing to avoid the subject, we'd sometimes argue politics, alone together in the cab of his truck for hours at a time. I probably annoyed him much more than he upset me. But I never missed a day, and I put out the work, so on the whole he was satisfied.

I did an eight-month construction season with Sneezy, and then he laid me off for the winter. Delighted to be free again, I applied for unemployment compensation and food stamps and was shocked when I actually got them. I was working under a phony name, but with a real Social Security number. Sue by now was pregnant again and had been advised by a doctor to take hormones and stay in bed for three months to avoid another miscarriage; I was able to stay at home and care for her during the pregnancy. All thanks to Sneezy.

The move from Santa Fe to Pennsylvania had brought us geographically closer to the organization, and several old comrades began making trips down to Pennsylvania from Eggplant houses in New York and Massachusetts to visit and keep in touch. A few even helped with work on the schoolhouse. It seemed as if we were a welcome break from their circumscribed, closed lives. Plus, to each other we were family, brothers and sisters who might have taken slightly different routes but were still close.

For the most part, my old comrades lived shadowy, anonymous existences and were in contact only with supporters who supplied them with money and information. People in the Eggplant were generally desperate to know what was happening in the movement, who was doing what, and how the latest communiqué or bombing was received. The contacts were mostly old friends who told the fugitives what they wanted to hear. This was especially true in the case of contacts of the organization's leaders, like Bernardine and Billy. We were regularly told, "People in the movement loved our latest action," whether referring to bombing the Pentagon, the Capitol building, or another site. "They thought it was a big boost to their work." Sue and I were not so sure.

This practice of telling people what they wanted to hear was not that dissimilar from the government's disastrous and corrupt practice of generating false information about "success" in the Vietnam War. Low-level officers would produce phony numbers of how many Vietnamese their units had killed, which would then be inflated at each command level up to the generals and the president. No one wanted to give a higher-up bad news.

I first became aware of this corruption back in the Weatherman collectives, when "second-level" cadres (those below the Weather Bureau) would tell me about how well their collective's work was going, how many people were being "organized," how many would be coming to the October National Action in Chicago. Inevitably the reports were untrue and often masked total failure. All hierarchies—based as they are on maximizing people's individual positions and egos—seem to have deception built in. So I generally distrusted and discounted the reports we were receiving secondhand about the underground's effectiveness.

Meanwhile, Sue and I began setting up our own network of contacts—old friends from New York City who would let us know independently what was happening in the movement. We, too, were hungry for information and old friendships, the movement having been our lives for so long.

After the 1970 Kent State killings, the student movement had declined nationwide, in part because the government shifted the draft to a lottery system and also because the numbers of recruits needed had diminished due to Vietnamization. And the shootings seemed to scare off students from demonstrating, which, after all, was what the government wanted to achieve. As student participation declined, returning Vietnam vets, organized into Vietnam Veterans Against the War, took up the front position in demonstrations and revived the entire antiwar movement. Vets regularly testified about the atrocities they had been forced to commit in Southeast Asia.

There were still occasional huge mobilizations—involving both students and nonstudents—such as the May Day demonstrations on May 1, 1971, in Washington, D.C., in which more than ten thousand people were rounded up by police and military in the biggest mass arrest in U.S. history. Thousands of young people expressed their opposition to Nixon's reelection at the demonstrations at the Republican National Convention in Miami in the summer of 1972.

My first meetings with old friends were inevitably anxiety-ridden: I had long looked forward to seeing these people, yet I worried about how they would react to me. Did they approve of my being a fugitive? What did they think of the organization and what it was accomplishing?

For the most part, we were overjoyed to see each other. Old friends were relieved to learn that I was still alive, while I was happy for the chance to renew friendships. After the initial catching up, I would always ask my questions about their perceptions of the organization. What I learned confirmed my suspicions: The Weather Underground didn't seem to affect anybody at all, in any way. We were not a part of most people's universe, even of those who were still working in what remained of the movement.

I became very cynical, wondering what all this effort was achieving. Sue revived her old name for the organization from 1970—the Bell Jar.

Our son, Paul, was born under an assumed name, Paul Joaquin Bowman, in July 1974, at Booth Memorial, a Salvation Army hospital in Philadelphia established in the nineteenth century as a home for unwed mothers; later it was opened to anyone. Sue had chosen the place because they used nurse-midwives instead of doctors. We had gone to Lamaze classes in Allentown, nearer our house, read books on natural childbirth, and practiced religiously every day at home. "Breathe, breathe," I'd say, massaging her, as Sue learned to pant and focus her attention as a conditioned response to my words and my touch. We adored the training for a natural childbirth, marveling at the fact that Lamaze had been a French Communist doctor at a clinic for metalworkers' families. Was there something in Lamaze's Marxist approach, we wondered, that had led him to develop a way for women to successfully overcome centuries of fearful cultural conditioning? He taught women techniques for coping with the pain of childbirth.

The labor turned out to be long and difficult, more than twenty-four hours. Sue hung in, even though she was wiped out. She never needed anesthetics. Playing my role of cheerful coach, I dutifully massaged her back, panted with her, talked with the nurse-midwives. Exhausted, Sue finally pushed Paul out, his head temporarily molded and swollen into the shape of a giant zucchini. We took turns holding him in the delivery room even before he was cleaned up, admiring him, dazzled by this strange and wonderful alien with perfect little fingers, ears, nose, the whole works.

Back then, in that humane Salvation Army hospital (long ago torn down to make way for a luxury condominium), mothers were encouraged to stay for three days, something almost unthinkable now. Sue began breast-feeding Paul, and we both learned from the nurses to hold and care for our baby without the stress of having to immediately change our environment. That period for me was a lot like being on an acid trip, with complete dissociation of mind, body, and place: I experienced moments of manic joy, sometimes followed by deepest fear. But there was our baby, an objective reality who needed our care, no matter our emotional moods, our legal situation, or our incompetence and lack of knowledge as parents.

We were all alone when we finally took Paul home in our ancient blue Chevy Nova, no grandmothers or relatives or even friends with kids we could ask for advice and help. I found it hard to believe that whoever was in charge (God? the government?) had entrusted him to us, since we knew so little about how to raise a baby.

The Watergate hearings were on television that week, and I carried Paul around the house for hours on end to comfort him and to quiet his crying. His lullaby was "Watergate, Watergate, we like our Watergate," which I sang to him over and over.

I like to think the Weather Underground had some small part in the 1973 and 1974 scandal called Watergate, which exposed Nixon as a crook, a liar, and a shredder of the U.S. Constitution, and which drove him from office. All federal charges against me and the other Weathermen were dismissed as Watergate unfolded.

In the fall of 1973, just as the Watergate story was gaining momentum, our attorneys in the big federal felony conspiracy cases from 1970 moved in court in Detroit to have the feds disclose how they had obtained the evidence against us. The judge upheld our motion, and the federal prosecutor, not wanting to admit illegality on the government's part, then dropped the charges against us, citing "national security." That was all I knew about it for some years. After I turned myself in, in 1977, I heard the full story from Gerry Lefcourt, my attorney.

The original Watergate burglary of the Democratic National Committee headquarters, in June 1972, which eventually would bring Nixon down, was only a tiny part of what Gerry and the other defense lawyers working on our cases referred to as " a criminal conspiracy" on the part of the government. For years the Nixon administration had been waging full-scale warfare against both the domestic opposition to the war and the black movement.

Starting in 1969, when Nixon first entered office, an Internal Security Division had been set up in the Justice Department, with ties directly upward to the White House and downward to the FBI and to state and local Red Squads (police investigative units). A series of prosecutions followed in the next three years: our conspiracy cases in Detroit and Chicago; the Pentagon Papers case around Daniel Ellsberg leaking to the *New York Times* an extremely damaging classified

Pentagon report on the history of the Vietnam War; the Chicago 8 case; an alleged plot by Catholic peace activists to kidnap Henry Kissinger; an alledged conspiracy to violently disrupt the Republican National Convention by some Vietnam veterans in Florida who were opposed to the war, known as the Gainesville 8; the indictments of the organizers of the May Day demonstrations in Washington, D.C., in May 1971, in which ten thousand were arrested; plus a whole series of Black Panther cases, such as New York's Panther 21. The government eventually lost every single one of these prosecutions, though they did succeed in tying up enormous amounts of the movement's money and energy.

In July 1970, Nixon had convened a meeting of the heads of the FBI, the CIA, the National Security Agency, and each of the military intelligence services to deal with what he perceived as the growing revolutionary threat of the Black Panther Party and the Weathermen. They took us at least as seriously as we took ourselves: Out of the meeting came a plan for a campaign of burglaries, wiretaps, infiltration of campus groups, use of military intelligence operatives, mail searches, and mail watches. Evidence of all these illegal means of gathering intelligence surfaced during the various Internal Security Division prosecutions of the mid-seventies.

In the FBI the campaign of political repression of dissidents had already become institutionalized long before Nixon, as COINTELPRO, the Counter Intelligence Program. It involved the use of infiltrators, provocateurs, wiretaps, break-ins, fabrication of evidence, disinformation, and harassment against the entire Left and antiwar opposition; the black civil-rights and revolutionary activists, including Martin Luther King Jr., Malcolm X, and the Black Panthers; as well as other minority activists, such as the American Indian Movement. Literally hundreds of individuals and groups whom FBI director J. Edgar Hoover deemed subversive were destroyed, jailed, or otherwise neutralized by completely illegal clandestine COINTELPRO activities. Fred Hampton was murdered by federal agents working with the local Chicago police.

In the course of preparing the defense of the Detroit case—officially known as *U.S. v. Rudd, et al.*—Gerry and the other attorneys uncovered evidence and depositions showing a host of illegal governmental activities against us:

- Three thousand pages of transcripts of unauthorized telephone wiretaps involving more than twelve thousand separate conversations. The Justice Department rationalized these wiretaps by claiming they were needed for national security, but the Supreme Court declared them illegal in June 1972.
- Grand juries impaneled in eight cities to gather evidence and launder illegally obtained evidence. Neither of these functions is considered a legal use of the grand jury. The original Detroit indictment was extremely vague and flimsy; it read, "conspiracy to cross state lines to blow things up."
- A series of kidnappings and beatings of relatives, friends, and acquaintances of Weathermen to gain information against us.
- Burglaries of homes and offices of people associated with the Weathermen. For example, the homes of at least five attorneys working on the defense of our Detroit case were broken in to between 1970 and 1972. In April 1970, right after the town-house explosion had forced us underground, the offices of the New York City Law Commune, which was also defending the Panther 21, were set on fire. Gerry Lefcourt was a member of the commune. After the fire was put out, the files on my cases were found open and strewn about.
- Mail watches. A second cousin of my father, who lived in Montreal, Canada, wrote a letter to my parents on May 5, 1970, saying that I would be welcome in his home if I got to Canada (I had just become a fugitive). Three days later, even before the letter was delivered to my folks, two agents of the Royal Canadian Mounted Police, part of which is the Canadian equivalent of the FBI, burst into our cousin's house screaming, "Is Mark Rudd here?"

The federal judge assigned to our cases in Detroit was Judge Damon J. Keith, a liberal black jurist appointed by Lyndon Johnson. Judge Keith had a special interest in civil liberties. In a 1972 decision, the Supreme Court upheld an earlier decision of his, in a case involving the revolutionary White Panther Party of Detroit, that wiretaps of domestic activists not warranted by court order could not be justified with the claim of national security.

Among the pretrial data concerning government illegalities piling up in our case was an obscure reference in a *Newsweek* article to the Weathermen's being a target of the White House "Plumbers," who had perpetrated the Watergate break-in. Particularly intrigued by that fact, Judge Keith ordered an evidentiary hearing for October 15, 1973, with a sweeping order demanding testimony from representatives of all federal agencies involved in intelligence gathering. A brave and principled man, he was curious to know if any illegal espionage had occurred and what, if any, was the tie to the still-unfolding Watergate case.

Gerry Lefcourt and our other attorneys scurried around the country gathering sworn affidavits from victims of government illegalities; they also formulated strategy to expose the government's witnesses. Even Tom Charles Huston, the former Justice Department attorney who was credited with writing the 1970 Intelligence Spying Plan (most often referred to as the "Huston Plan") offered to cooperate with our attorneys.

Meanwhile, the federal law-enforcement bureaucracy was in a state of disarray. As a result of Watergate, a number of heads had started to roll: John Mitchell, under indictment now for his role in the cover-up, was no longer attorney general; Robert Mardian, also indicted, was out as head of the Internal Security Division of the Justice Department; L. Patrick Gray, who was serving as acting head of the FBI, was forced to resign under a cloud of scandal; Vice President Spiro Agnew, who had served as Nixon's attack dog against students and other opponents of the war, resigned and pleaded no contest to corruption charges just five days before our hearing. And Nixon himself was being threatened with impeachment. The government was in no shape to face the disclosures that our lawyers were planning for the hearing.

So the feds moved in court to drop all charges, which Judge Keith did "with prejudice," which meant the case could never be revived. The Justice Department attorneys, true to form, cited national security as the reason they wouldn't disclose how their evidence had been obtained.

The dropping of all the federal charges had not the slightest impact on my life as a fugitive. I assumed that the feds were still looking for us, since the organization had committed many bombings in the intervening years. But there was a more important reason to stay underground: If I surrendered, it would have burst the myth of righteous guerrilla fighters surviving in the belly of the imperialist beast, despite the best efforts of the supposedly all-powerful state security police. My name remained a symbol for the revolutionary underground, no matter how far removed I actually was. I continued to be loyal to the organization and its purpose, so maintaining the myth seemed a useful function in itself, even though I was accomplishing nothing else. After all, imperialism still existed, and the Vietnam War was still raging, financed by U.S. government money.

By the fall of 1973, also, the Weather underground had restructured itself into the Weather Underground Organization (WUO), meaning it had been rebuilt with a leadership body known as a Central Committee, cadre or members organized into collectives, and a political program. Sue and I heard rumors of this "consolidation" from our occasional visitors.

In addition to the fugitives, a network of aboveground members was being organized formally, also into secret collectives. This dual approach—involving both fugitives and nonfugitives—mirrored the organization of all successful Communist-type parties, especially the National Liberation Front in South Vietnam. We in the underground were referred to as "the forest" in WUO code; the aboveground members and supporters were "the ocean."

Besides setting up this structure, the main work of the consolidation was the writing and printing of a political statement, a work that would express the WUO's view, quite literally, of the whole world. It took more than a year—from late winter 1972 to the spring of 1974—of drafts, discussions, revisions, for the newly resurrected organization to produce a full-fledged book, *Prairie Fire: The Politics of Revolutionary Anti-Imperialism.* The difficulties involved were only in part a result of its clandestine authorship: The book attempted to be encyclopedic. It included sections on the history and development of the revolutionary movement in the United States (complete with self-criticism and criticism of others); the recent withdrawal of American troops from Vietnam and the importance of maintaining the antiwar movement; a rendition of U.S. history focusing on the oppression and resistance of Native Americans, blacks, women, and workers; an analysis of U.S. imperialism in the Third World, with accounts of ongoing struggles for national liberation in Puerto Rico, the Portuguese African colonies, and Palestine; a class analysis of the home front, including sections on the struggles of black people, Native Americans, Chicanos, women, and youth; and, finally, a call for a mass revolutionary movement based in internationalism and militancy.

The only role I played in the writing of *Prairie Fire* was to read a section of a draft someone might occasionally show me. I was impressed by the enormous effort involved, not just in the writing but also in the production and distribution of the work. Everything had to be done secretly. A complete hidden print shop was set up; paper and supplies were surreptitiously obtained; the pages were printed, collated, bound, and mailed—all in such a way as to leave no trace of origin. The people handling the book had to wear gloves at all times—no fingerprints. About a thousand copies were mailed out to leftist organizations, bookstores, newspapers, magazines. In New York and San Francisco, aboveground supporters created organizations known as Prairie Fire Distribution Committees to reprint and sell the book. Within two years thirty-five thousand were produced and distributed by these secondary means.

I have no way of knowing the true impact of *Prairie Fire.* It did not revive the Left, as was hoped; nor did it build a Communist Party in the United States, which was the stated aim. Contacts of mine in and around the movement spoke favorably of the manifesto, and I did get a sense that its existence let many leftists believe that the underground was still alive.

The book was an attempt to influence the movement that we had abandoned back in 1969. The WUO tried to reach out to the many thousands of New Leftists and former New Leftists by saying, in effect, "Don't despair, we're all part of the same thing," but at the same time the organization unwaveringly maintained anti-imperialism and armed struggle as principles. *Prairie Fire*'s overall tone was omniscient to the point of arrogance. There was also, of course, the obvious irony that a book, something entirely legal, was the hard-fought product of an elaborate and costly clandestine organization. If anybody stopped to think about it, that made no sense at all.

My own feelings toward my comrades in the WUO became increasingly ambivalent during the winter of 1973 and spring of 1974. I respected their energy and also needed them as friends and support. On the other hand, Sue and I both disagreed with the WUO's position on the Symbionese Liberation Army (SLA) and the kidnapping of Patty Hearst.

We thought the whole business was dangerous and destructive. Despite calling themselves "revolutionaries," the SLA were true terrorists, without any limits or any sense. They claimed they were acting for the liberation of black people, but actions such as the assassination of Marcus Foster, the first black Oakland school superintendent, or their spraying with bullets a bank lobby filled with customers could only be interpreted as terroristic. Also, I never believed a minute in

Patty Hearst's conversion from self-centered apolitical young heiress to guerrilla fighter code-named "Tania."

The WUO supported the SLA in public statements; they even published ridiculous poems in "revolutionary solidarity" with the organization, which was centered on one black man, an ex-con, and his small coterie of white followers. I was told that Billy Ayers and Bernardine Dohrn stood in line and talked to people receiving free food financed by the Hearst family at a distribution center in California. The food distribution was one of the ransom demands for Patty's release. In Billy and Bernardine's view, the positive comments they're alleged to have heard from the other food recipients—that the food distribution supposedly highlighted the existence of hunger in America—justified the whole business.

The day in May 1974 that the Los Angeles police murdered six members of the SLA, including Cinque, the leader, but not Patty Hearst, two old friends from the Eggplant happened to be visiting us at the schoolhouse. Together we watched film clips of the police siege, during which they fired nine hundred rounds into the house before setting it on fire.

"The pigs are murdering them!" one of my friends exclaimed. "Now everyone will see how righteous the SLA are. They're burning them alive."

"I don't think so," I said. "This is just a cops-and-robbers TV show, where the bad guys die and the cops win in the end."

"Can't you see they're revolutionary martyrs?" he replied.

"I don't think they're anything more than self-deluded. Most normal people probably see them as criminals," I shot back.

"You're totally right-wing," was the final word on me and my views on the SLA.

On May 31, 1974, the Weather Underground Organization retaliated for the murder of the SLA 6 by bombing the Los Angeles office of the California state attorney general. Patty Hearst was eventually captured in September 1975. At her trial the next year, she was no longer the revolutionary guerrilla Tania and was back under her billionaire father's control. Hearst pleaded innocent on the grounds of Stockholm syndrome, a psychological condition in which captives have been known to sympathize with their captors. Hearst also testified against her surviving ex-comrades. It was to no avail, as she was sentenced to seven years in jail. (She got out after twenty-two months.) She then married her bodyguard and eventually took up a part-time career as a no-talent B-movie actress.

In November 1974, Sue and I received some disastrous news: Jane Alpert, whom we'd befriended back in Santa Fe two and a half years before, had turned herself in to the FBI. She'd hired a high-priced former New York City prosecutor with the best connections in the judicial system to pull strings in order to get her a short sentence. Our first reaction was that she was going to rat on us.

In a way she already had.

The year before, in the summer, she had published a long, two-part essay in *Ms.* magazine, the leading feminist publication of the time. In the second part, a theoretical treatise called "Mother Right," Alpert propounded the thesis that women's reproductive biology is the basis for their power, which she said is denied under the current patriarchy. This abstract theory was really only cover for her principal point: that she was breaking with her own past and "the male-dominated Left." This she accomplished in the first part, an open letter addressed "To the Sisters in the Weather Underground."

In this letter Alpert attempted to convince the Weatherwomen to abandon the Left and switch over to feminism, as if the two were totally incompatible. Her proof that the Left was merely part of the oppressor patriarchy consisted of an anecdote concerning the sexual crimes of a certain well-known New Left leader in the Weather ranks: me. She recounted—in a garbled version, but with intimate details—a confession I had made to her in Santa Fe while we were discussing feminism and women's liberation.

In California back in 1971, as I was leaving the organization and deciding whether to hook up with Sue, I had become infatuated with a married woman whose house I was staying at. We'd had an affair. The woman became pregnant and had an abortion. The whole business had hurt Sue deeply, and I was ashamed. Sue and I were both mortified to see this story exposed in a national magazine. But worse, the piece gave the FBI a big gift of information—that I had a woman partner underground. While claiming that her aim was to save women in the underground from the oppression of sexist men like me, Alpert had betrayed Sue by exposing her to the government.

For months afterward, each time I unloaded a forty-pound concrete copingstone off Sneezy's truck, I imagined placing Alpert's miserable little head under one of the blocks and smashing it to a bloody pulp. That cheery thought kept the hours moving right along.

In time my anger burned itself out, but eighteen months later, just three days after Alpert surrendered, I received word from Santa Fe that the town had been flooded with FBI agents looking for Sue and me. An article in the local daily newspaper, the *New Mexican,* stated that two Weatherman fugitives had lived in Santa Fe in 1972 and gave the names we used. It was clearly an attempt to flush out information. Agents visited one of my ex-bosses, the hippie contractor, and asked for information about Tony Goodman. Scared to death, and wanting to please them, he ran to pull out a group photo of the crew I had worked on. There we all were, but Tony Goodman, in the back row, had his face covered with a bandanna, looking like a *bandido* not very happy to have his picture taken.

My former neighbor, John, a carpenter, was grabbed by FBI agents, put into a car, punched around briefly, then threatened with arrest for having dealt marijuana to Jane Alpert, which was a total fabrication. Actually, he'd never met her. The agents showed John a picture of himself and me at the J. S. Horn lecture in the church where I'd first encountered Alpert. Apparently the local FBI office routinely took pictures of everyone attending "subversive" events such as a lecture by a doctor who had practiced in China.

Sue and I were only one jump removed from Santa Fe. Though

we'd been in our renovated schoolhouse for two and a half years, longer than we'd lived anywhere else, and sorely hated to leave it, we could find no alternative but to move once again. Especially now that we had a baby, we needed to take every precaution.

We later found out that we were right to have feared being busted. In 1975, after we'd moved, we learned from friends of friends that a description of one of the dogs we had acquired in Santa Fe—Ernie, a striking quarter timber wolf, three-quarter malamute—had been circulated to vets in the New England area. Also, the pictures from our New Mexico licenses were sent out to health-food and natural-food stores in that area.

I had been tempted after the federal charges were dropped in 1973 to believe that the active search for us had abated. Guy Goodwin's grand juries, for one thing, had stopped: Watergate seemed to have disorganized the secret police apparatus. But the details for the search for us began to drift in at the same time as the search for Patty Hearst reached hysterical proportions. At one point Hearst was reportedly spotted in eastern Pennsylvania, and I began to wonder if it was safe to go out of our house.

In the two years following Alpert's surrender, a debate raged aboveground in the Left and in the women's movement over whether she had snitched on those of us still under. People lined up on each side of the issue on completely ideological grounds, with little evidence one way or the other. Those who agreed with Alpert that the Left was antiwoman defended her; those who believed in the primacy of anti-imperialism and antiracism, or just refused to accept a hierarchy of oppression that Alpert imposed, attacked her.

I ardently wanted the WUO to go public with my information from Santa Fe. *Why not hit her back if we could?* I was quite open about my motive—the sweet revenge of exposing and isolating this person I believed to be a traitor.

The women of the organization—led by Bernardine, but including many other old friends—would not hear of it. They claimed that the information would jeopardize Sue and my contacts in Santa Fe. That was bunk, because the FBI had gone public at the outset by planting the article about the search for us in the Santa Fe newspaper. Their real reason for not wanting to counterattack Alpert was that the women in the organization felt vulnerable to her charges of being lousy feminists in a male-dominated organization. Weatherman's whole existence was macho, especially trying to prove ourselves through violence. The last thing they wanted to do was associate themselves publicly with a sexist criminal like me.

I felt betrayed down to the core. What happened to friendship and loyalty? Sometimes I didn't know whom I hated more, Jane Alpert or the women in the organization who failed to rise to my defense. The hurt took years to dissipate.

Eventually I grew to pity Alpert, especially after she published a sad autobiography in 1981, *Growing Up Underground*. She pathetically described how she had been dominated first by Sam Melville, her lover who was responsible for her becoming a bomber and a fugitive, and then overwhelmed by the brilliance of Robin Morgan, a noted feminist poet and author. Morgan, famous for, among other things, an influential 1970 anti-male-leftist screed titled "Goodbye to All That," had fed Alpert the separatist anti-left ideology that became the *Ms.* article. She also helped get it published there (later becoming editor of *Ms.* herself). And it was she who found Alpert the well-connected establishment attorney who orchestrated her surrender. Alpert wound up serving two years' prison time, after which her friend Robin Morgan dumped her.

16

WUO Split

Victor and Irene Kelso lived with their toddler son, Paul, in the second-floor, two-bedroom apartment in a downtown working-class enclave of New Rochelle, a wealthy suburb of New York City. Victor could often be seen pushing Paul around in his collapsible stroller on their shopping rounds, to the public library or the park, or, in the summer, to the little town beach looking out on the western end of Long Island Sound. Irene worked as an aide in the neighborhood Head Start day-care program. She was well liked at the job, funny and caring with the kids, and a good co-worker. Victor was a graduate student at Fordham University in the Bronx but had arranged things so that he was home with their son when Irene was at work. Everything was cool.

The toddler in this little story was indeed my son, Paul, but "Irene" and "Victor" were only our most recent identities, hastily adopted when we fled Pennsylvania right before Christmas 1974, when Paul was just five months old. We needed to put one additional step between ourselves and the last place we were now known to have been living, Santa Fe; we had chosen to live on the outskirts of New York City in order to be closer to the Eggplant people. They had two apartments, one in the west Bronx, the other in Yonkers. Also, we wanted to be closer to our own old friends from New York City, with whom we had begun to be back in touch. Now that we had a kid, we didn't want to be so alone out in the world, and we longed to be with people who knew us as we really were.

From the time Paul was born, Sue and I had begun to reassess what we were doing with our lives. More than anything else, being a parent increased my ties to the "normal" world. I began to understand the dilemmas and pressures that all parents feel: what values to transmit within the family, how to protect and support your kid, how to raise him or her in a decent community, with friends and family. Plus, I now felt close to my own parents in a new way. Every time I'd wipe Paul's bottom, I'd reflect on the two people who had wiped mine twenty-seven years before and how much I owed them. When would Paul know his grandparents?

Sue's job in the day-care center did solve the problem of meaningful work. For years she had taken any job that came along—secretarial, retail, what have you—but she needed something more fulfilling that would let her use her mind and her many skills. So she volunteered at the day-care center near our house, and just as we had hoped, when an opening came up, the women running the program hired her. They knew a good worker when they saw one. Her job became a joy for her; it was the first she had any interest in during all her years underground with me.

I spent 1976 as a househusband—I was not a graduate student, that was only a cover story—taking care of Paul, cleaning, shopping, cooking. To some extent I suffered from the same malady that afflicts many housewives—lack of self-esteem, loss of confidence in my ability to deal with the world. Though I knew better, I didn't honor my housework and child care as real work.

On the other hand, that year allowed Paul and me to grow intensely close to each other. There were no "terrible twos" as far as I was concerned. We'd spend every day playing and reading together, taking walks, doing errands, napping. That was a glorious, luxurious way to live. Paul was an especially sunny and loving kid. He was a late talker, a kid who picked things up by watching and thinking. We counted everything—shoes, train cars, animals—and he caught on so fast that he could soon count infinitely in both directions, though he didn't pick up negative numbers until he was about five. I am privileged to have had the opportunity so few men have, to bond at this age with their baby. Had I not been a fugitive, maybe it wouldn't have happened that way.

The WUO had won such a great "victory" with *Prairie Fire* that the leadership decided that next the organization would put out a quarterly magazine to broadcast the underground's ideas to the movement. Until the creation of *Prairie Fire,* an occasional bombing accompanied by a communiqué was the only way to inform people of what the WUO was thinking.

Osawatomie, the magazine of the Weather Underground Organization, first appeared in March 1975. It was named after John Brown, the Great Liberator of the pre–Civil War era, who had been given the nickname following a battle at Osawatomie, Kansas, in 1856. Brown and thirty others defeated a much larger pro-slavery force bent on making Kansas a slave state. To the Eggplant people, including myself, Osawatomie was a powerful historical example of white people fighting for the freedom of blacks. Malcolm X once said, "If a white person wants to help our cause, ask him what he thinks of John Brown. Do you know what Brown did? He went to war."

The magazine was slick, given that it was homemade. It even had a color cover. The first issue's main feature story was about the Boston school busing crisis. The article revealed how WUO members had infiltrated the leading anti-busing group, ROAR (for Restore Our Alienated Rights), in order to "gather knowledge of the enemy."

Despite its claims of merely defending local schools against court-ordered busing, ROAR was virulently racist, the magazine reported.

I witnessed a white riot in Boston myself in the spring of 1975. I was staying at a Weather apartment in Chelsea, one of the neighborhoods affected by busing. Early one morning I was on my way to the subway when a mob of white teenagers and a few adults attacked a school bus filled with black kids. I was close enough to see the terror on the kids' faces when the white crowd broke windows, forced a door open, and dragged a black girl off the bus and beat her before running away. It all happened so quickly that I saw no chance to intervene; I was shaken. At least we had a magazine to expose the racism, I thought.

I was at that time temporarily in Boston in order to help build a new, expanded print shop, complete with a photo darkroom and plate-making facilities. Working with another fugitive who was a master carpenter—much more skilled than I—we converted an old loft across the street from the Boston Gardens into our secret print shop. We even built an entry area with a reception desk, to stop occasional visitors from poking about where the work was under way. The noise of the press was impossible to mask, so we convinced the landlord and the other tenants of the building that we were a small social-research and consulting business that had its own press.

Osawatomie continued to come out quarterly until April 1976, when it became a bimonthly. The "Bicentennial Issue," dated June–July 1976, was destined to be its last, however. The cover featured an amazing photo of a white riot in front of Boston's city hall. A white guy was spearing a black man, whose hands were locked behind his back, with a pole carrying an American flag. The caption read, "200 Years Is More Than Enough."

As with *Prairie Fire,* I have no idea how important *Osawatomie* was in the real world. It was mailed in bundles of ten or twenty to leftist bookstores and publications, and also reprinted by our aboveground supporters, the Prairie Fire Organizing Committees (PFOC). It might have been read by several thousand people in total, all already loyal to our point of view. Probably the main effect was internal, as it gave WUO fugitives a way to clarify their ideas and also to develop a unified organization. I doubt whether it successfully broadcast those ideas very far beyond the Bell Jar, however.

But a much more effective medium of communication presented itself. Emile de Antonio, a well-known documentary filmmaker, approached the WUO through a friend of mine with the idea of doing a film version of *Prairie Fire.* De Antonio had made *Point of Order,* about the rise and fall of Senator Joseph McCarthy, the red-baiter, and *In the Year of the Pig,* about the U.S. invasion of Vietnam. He certainly had the necessary left-wing credentials. The WUO Central Committee jumped at the chance to be movie stars; this was consistent with the new trend in the organization away from bombings and toward media work.

The movie that resulted, titled *Underground,* was first screened for the members of the New York City forest, the actual underground, in Sue's and my kitchen in New Rochelle. Most of the movie consisted of profiles of five Weather fugitives shot through a screen to mask their appearance. Behind the scrim, Bernardine, Billy, Jeff Jones, and the town-house survivors, Kathy Boudin and Cathy Wilkerson, went on endlessly about the reasons for going underground, the justifications for violent revolution, and how they lived. The movie was so talky that its code name in the organization immediately became *Jaws,* a name coined by Sue and me. The only relief from the torrent of verbiage was an occasional bit of stock footage to illustrate a point. Some of it was quite dramatic, such as the helicopters being pushed off the side of the aircraft carriers during the U.S. evacuation of Saigon in April 1975.

Much of the talk played up the heroism of being a Weatherman, which I felt was a self-serving distortion of reality. Life underground was mostly insanely boring, the low-grade daily anxiety more like a dull ache than what one sees in a TV adventure movie. On a physical level, the underground was almost cushy; friends and relatives and political supporters supplied most of the money.

I felt that the whole movie project was irrelevant. The underground had ceased to be a significant factor even to the Left, and after the 1973 cease-fire in Vietnam and the final victory by the Vietnamese in 1975, the activist Left itself had shrunk to deluded remnants fighting over Talmudic points of religious difference, such as whether the greater threat to American workers was U.S. monopoly imperialism or Soviet social imperialism. Even had *Underground* been a better film, it could hardly have resurrected a corpse.

The movie did achieve a measure of political significance, however, thanks to the predictably heavy-handed blundering of the U.S. government. The FBI in Los Angeles heard about the film, probably because several well-known movie people were involved in the work, including Haskell Wexler, the Academy Award–winning cinematographer. As a result, de Antonio, Wexler, and Mary Lampson, the editor, were first harassed—their homes and workplaces broken into—and then they were subpoenaed before a federal grand jury while the film was in its final editing phase. Apparently the feds were still hot to find us. Instead of meekly surrendering to the government's intimidation, the three went on the offensive and organized a large support movement, involving some of Hollywood's top actors and directors of the time—Warren Beatty, Jack Nicholson, Sally Field, Jon Voight, Harry Belafonte, Mel Brooks, William Friedkin, Arthur Penn, and Elia Kazan, among others. The American Civil Liberties Union also joined in. This spirited defense proved to be the exact opposite of the Hollywood 10 case of the late forties, in which the terrified film industry (and the ACLU) capitulated for the most part to the government's Communist witch-hunt.

In a much-publicized news conference, the three filmmakers and their supporters charged that the feds wanted to suppress the film and also use the grand jury as a fishing expedition to find the Weather fugitives. They threatened to file an injunction to stop the government from interfering with their work. The feds backed down, the subpoenas were dropped, and "the film that the FBI didn't want you to see" opened in late 1975 to a chorus of yawns. It was shown mostly in college-town art theaters and also on television in Europe.

Underground, Prairie Fire, and *Osawatomie* were all part of the WUO's attempt to put itself forward, from below, as the leadership of a reconstructed Left, a fantasy based on their conception of how successful revolutionary parties were organized in Cuba, Vietnam, and China. Capping the effort, the leadership used its aboveground cadres in the Prairie Fire Organizing Committee to pull together a huge mass gathering known as the Hard Times Conference, held in Chicago in January 1976. Amazingly, more than two thousand people attended, proof of the large number of activists who were looking for organization that would revive the mass movement of the late sixties and early seventies.

At the conference PFOC pushed resolutions calling for "working-class unity" as the basis for combating the economic depression that had gripped the country since 1973. Whatever this might mean in practice, the phrase was a leftist code that angered the large number of blacks, Latinos, and feminists present, each group fearing that their separate demands and needs would be suppressed. They formed themselves into caucuses and angrily denounced the Hard Times organizers, especially PFOC. The conference broke up in disarray.

It was the old debate that had split SDS back at the 1969 convention, when the SDS regulars led by Weatherman took the side of Third World national liberation, while antinationalist Progressive Labor proclaimed "working class unity." Only now the WUO had switched sides and adopted PL's Marxist-fundamentalist class line. Nonwhite people and women at the conference were outraged.

The WUO leaders, searching for a theory and a blueprint to guide them, had been drawn back to the fundamental Marxist texts such as *Das Kapital.* What I think happened was that Eleanor Stein and Jeff Jones had recruited Eleanor's mother, Annie, an old Communist who'd come up during the Stalin era, as a theoretical adviser to the organization. Annie brought her volumes of Marx, Engels, and Lenin with her, and she must have come to the conclusion that the Eggplant, in *Prairie Fire,* had missed the correct magic Marxist formula. In a pre–Hard

Times Conference editorial in *Osawatomie,* the Central Committee confessed to having "ignored the *historic mission* of the working class and failed to energetically pursue our task of forging it into a conscious class, prepared to fight for the interests of the class as a whole." That sentence could have been written in 1905 or 1917 by Lenin himself.

The WUO's high ride on classical Marxism didn't last long. Not only did the nonwhite people and the feminists at the Hard Times Conference angrily turn against them, but many PFOC members, especially the majority who were not clandestine members of the WUO, also felt that they had been used. They claimed not to have known that there was a hidden cabal—the WUO leadership—pulling their strings from below. Within weeks of the conference, a "left-wing" faction emerged in the WUO, led by none other than Clayton Van Lydegraf, the old Maoist with whom I'd worked on the Leary action back in 1970.

Van Lydegraf, a loyal cadre of the WUO, had gone aboveground to help found PFOC around 1974. But now he attacked the WUO Central Committee for abandoning the organization's historic position on the primacy of fighting imperialism. To Van and his followers, the WUO had committed the classic error of "opportunism," taking the easy road of not dealing with racism and national liberation in order to gain access to the workers.

Horrified, the leadership of the WUO immediately backtracked and attempted to recant their errors in a process known internally as "rectification." The last issue of *Osawatomie* contains a long, abject self-criticism signed by the "Central Committee, WUO." But the formal confession wasn't enough. Tasting blood and seeking revenge for several personal slights he had suffered at the hands of Bernardine and her clique, Van Lydegraf kept up the attack.

All this infighting was beyond absurd, of course. Neither the WUO Central Committee nor the Revolutionary Committee, as Van Lydegraf's group came to call itself, had the slightest following among workers or anyone else. They were like madmen arguing about what they'd do with an imaginary inheritance. Even more ludicrous, at the core of the rectification was a battle between two ancient Stalinists from the Old Left, Van and Annie, brutally slugging it out over issues that they'd been contending for thirty or more years.

Believe it or not, exactly at this moment, as things were coming apart, I chose to try to try to get readmitted to the organization I'd left back in 1970. I had no interest whatsoever in the ideological split around the breakup of the Hard Times Conference. My goal was simply to figure out the best way to emerge from the underground. With a two-year-old, Sue and I could see that when Paul reached a more social age—when he was talking and going to school—we'd need to stabilize our situation, for his sake and ours. Also, I missed my family and wanted them to get to know Paul. We longed to be real people in a real community, not anonymous shadows, which is what we were in New Rochelle.

As for the tired effort of maintaining the myth of the underground, the war was now over; the Vietnamese had won on April 30, 1975. Imperialism still existed, but there was not going to be a revolution in this country in any foreseeable future. Especially after the debacle of the SLA, being underground seemed a pointless and needless sacrifice. We had come to the bitter realization that rather than doing any useful political work, we were just surviving. By early 1976, Sue and I resolved to try to come up, one way or another.

By attempting to rejoin the organization at a moment of crisis, I thought I'd be able to convince my fellow fugitives to bring the whole organization aboveground. I hoped that there would be safety in numbers. Though the heavy federal charges had been dropped back in 1973, I had no way of knowing what additional charges would be brought against me or anyone else who surfaced. The WUO had publicly taken credit for more than twenty-four bombings; plus, the three deaths in the town house had to be legally accounted for. The feds were still actively looking for us.

The battle between the Central Committee and the Revolutionary Committee had weakened the leadership. That was fine with me: I had little sympathy for Bernardine, Bill, and Jeff at this point. They had overreached themselves to my mind, with arrogant claims to "revolutionary leadership" for themselves and the organization. Our strategy of going underground in 1970 had simply been the wrong choice.

My part in the destruction of the Weather Underground was actually very small. I did manage to advance the idea of "inversion"—that is, bringing the belowground up out of hiding as one big unit. The Central Committee even jumped onto it for a brief moment, seeing the possibility of offering their leadership to the mass movement, which is what they'd been trying to do by remote control for the last two years at least. That was not what I had in mind.

To the Revolutionary Committee, this new line around inversion was red meat. They considered surfacing the organization to be total capitulation to imperialism, because it would surrender the capacity for armed struggle. (A year later Van Lydegraf and associates, in their zeal for "the highest stage of struggle," were busted in a bomb plot that was infiltrated by a police agent.) By the early fall of 1976, Lydegraf's group had assumed leadership over the whole deluded organization, dethroning Bernardine's old Central Committee. The first thing they did was to order that no one could even *talk* to anyone else who had been in the old leadership or been a supporter of theirs. I was shunned by everyone on both sides as a complete right-winger who had proposed inversion.

It hurt deeply when several old friends, people I loved dearly, refused to see me and Sue because of this bizarre prohibition. Sue and I viewed this mass shunning as paranoid, psychotic, and manipulative, quite worthy of anything Stalin had pulled off, minus the firing squads.

Continuing in the Stalinist vein, the Van Lydegraf's Revolutionary Committee released a pathetic tape, played at several public events, of Bernardine abjectly confessing her crimes against the people of the world and the revolution. She copped to, among other ridiculous charges: "naked white supremacy, white superiority, and chauvinist arrogance," and to "denying support to Third World liberation." Additionally, she confessed that "in 1974, we set out to destroy the women's movement." She even named names of her co-conspirators—Jeff Jones and Bill Ayers being the leading criminals whom she denounced. One wonders what forms of mental anguish she went through to accede to making such a disgusting tape.

By Thanksgiving of 1976, Sue and I decided to begin the process of coming up on our own. The organization no longer existed, and a mass inversion seemed unlikely. The effort had degenerated into mindless Stalinism, cruelty, and betrayal. Even if I had to do time in prison, I thought, it would be better than this madness.

As if guided by some mischievous hand of destiny, who should come hulking back into our lives at this precise moment, the end of 1976, but JJ, John Jacobs, my old buddy sent into exile six years before as the sole living scapegoat for the town house. Smelling a chance, he had come back to check out what was happening with the organization and also to try to influence "the internal struggle" to go in his direction. He wanted to recruit me, his old sidekick, of course.

"Now that Bernardine and Billy and Jeff have fallen from leadership," he told me, "we can take over and rebuild the underground."

A revolutionary Peter Pan, JJ was still attached to the fantasy of a secret guerrilla army fighting the state, even though he had done nothing more than hide out for six years. His thinking was stuck back in 1970. This time I wasn't jumping for his bait. "Sue and I are going up. We're just wasting our time here," I rebuffed him.

"Why? What will you do aboveground—if you're not in jail? Join the Democratic Party?"

"Any form of organizing is better than this," I replied. "At least we'll be accomplishing something."

JJ just shook his head, disgusted. I had become a right-winger on the order of William F. Buckley.

JJ and his partner, a beautiful young Canadian woman, and their newborn baby girl moved in with us in our second-floor apartment. Sue and I made up a cover story for our downstairs neighbors, who were relatives of the landlord—a vague but believable tale about my cousin visiting from California. How long they were staying remained open. I sensed they had no place to go and would camp with us forever, simply out of inertia. The comings and goings increased our visibility in the neighborhood and lowered our security; people might begin to ask more questions about Vic and Irene.

One evening JJ arranged for me to pick him up at the end of the subway line in the Bronx, about twenty minutes from our house. He was in the city seeing a contact. I got his call after midnight and quietly left the apartment, not wanting to wake the neighbors downstairs in the old frame house. On the way back from the Bronx, JJ and I got into one of our habitual overheated discussions about the future of the underground and the revolution. With the neighbors still in mind, we parked outside for a few minutes and continued to argue.

Suddenly a police cruiser pulled up next to us, and the lone cop, a white guy in his early twenties, shone his light in the driver's-side window.

"What are you doing here?" he demanded.

"We're just visiting our friend who lives in this apartment," I replied.

"Let me see your license and registration. His ID, too," said the cop, gesturing at JJ.

We turned over our licenses and the car registration.

Sue and I made it our iron practice never to link our car registration or our drivers' licenses to the place where we lived. We each had two sets of ID, one for the household and money matters such as paychecks, the other for driving. In addition, the car was registered to an entirely different name and address than either of our drivers' licenses. The system was cumbersome yet necessary, because it insulated the car, which could always be disposed of, from our house and our jobs, which were more difficult to abandon. The separation was in general an advantage, except now I'd been stopped by a cop at 1:00 A.M. outside my apartment, carrying ID papers completely different from the identity I lived under.

After the cop called in a check on our IDs, he told us to get out of the car. He directed me to go to the front and JJ to wait at the back. He then asked me who my friend was. Fortunately, I had memorized JJ's fake name and address. I sure hoped JJ would be as successful with mine. Things tend to slow down at moments like this; every move is fraught with disastrous implications, and you try to anticipate what will happen next. Yet there's no real planning possible, because you're just reacting instantaneously to the real lead in this drama, the young policeman who unknowingly holds your fate in his hands.

JJ answered his questions about our IDs successfully. Something was still bothering the cop, though. "What are you doing here?" he asked us.

"We came down from Connecticut to visit our friend Victor Kelso, and we didn't want to wake him and his wife and kid," I said, talking about my other self.

"Do they know you're here? How often do you visit?"

"About once a month," I said, completely winging it.

"Which apartment is it?"

"Upstairs. The door's on the left side of that porch," I instantly replied.

"Ring the bell, let's see if they know you."

I was curious myself to see how this was going to turn out, especially because we'd be waking up the neighbors, who were definitely not in on the joke.

I rang the doorbell.

Sue answered from the top of the stairs in her nightgown, "Who is it?"

I tried the door: I had left it unlocked. "It's me, Irene. There's an officer down here wants to know if you know us."

Sue peered down the stairs, wide-eyed, at the three of us crowding into the doorway.

"It's okay, Officer, that's our friend John Brower from Hartford, Connecticut. The other guy is Wayne, his friend."

The cop looked at her, still slightly suspicious.

He asked, "How often do you see this guy?"

My heart stopped.

Sue replied, "About once a month."

The young cop looked at us, looked at Sue, and put his hand to the brim of his hat in that timeworn salute.

"Sorry to get you out of bed, ma'am, but I was just doing my job."

"That's okay, Officer. Come on upstairs, you two goons. What are you doing waking us up in the middle of the night?"

Sue and I stayed up the rest of the night trying to figure out the implications of what had just happened. In the short run, we were safe. But how long until that cop would see me in the neighborhood and begin to ask questions about "John Brower" and find out I was better known to the neighbors as "Victor Kelso"? Also, the car was now tied to our house, at least in the officer's notebook. We had a shaky situation.

Worse, JJ had been using his only set of ID, despite our having asked him to have two, one under which he lived, another for traveling. I suppose that because of being separated from the organization for so many years, and therefore having fewer resources to call on, he hadn't tried to build a second set, which was a tedious and frustrating job. Also, JJ was naturally lazy, always had been.

I was planning to turn myself in within a few months, so there was a risk that the cop might see my picture in the paper, recognize me even with a disguise change, and then check his notebook and find my ID information and JJ's. Then the FBI could just call the Royal Canadian Mounted Police in Vancouver, Canada, and tell them to go and pick up JJ and family. JJ would need to abandon his house and identity. He and his partner were not pleased when we told them about the potential danger.

That night Sue and I resolved to move as soon as feasible, then wait a while before resurfacing, possibly up to a year, in order to lessen the chances of an accidental link to JJ. This would also give him time to change his ID and place of residence.

There it loomed again, the fourth major move since Sue and I had been underground together, this time under tight conditions and with a two-year-old toddler. Sue would have to give up her beloved work in the day-care center, a job she'd maneuvered for over a year to get. We would have to tell lies to our neighbors about our leaving, find a new apartment in a new city, start over again with new jobs, new friends, new ID, all because of a tiny security screwup.

How much longer would we have to keep doing this?

I never saw JJ again after that episode at the end of 1976. He returned to Vancouver and resumed his life as "Wayne Curry," beloved hippie-community raconteur and low-level marijuana dealer. I'm told that with every bag of pot he sold, he dispensed a lecture on U.S. imperialism. He helped raise two kids and numerous nieces and nephews, who still love him dearly. A niece told me, "Wayne raised us as proletarians." He died of skin cancer in 1997, having remained a Weather fugitive for twenty-seven years. JJ was not one to surrender. The principal author of the original Weatherman paper, JJ was both the first Weatherman and the last.

This was the man who had taught me anti-imperialism. My feelings about my old running partner are a confused blend of love and gratitude, anger and blame, spiced with a dash of competitiveness, which I'm still sorting out, ten years after his death.

17

A Middle-Class Hero

On September 14, 1977, at 6:00 A.M., I was sitting in the #1 Broadway local on the New York City subway, my old train from Columbia days, racing from the elevated Bronx down to deepest lower Manhattan. It was way too early for my 8:00 A.M. appointment with my attorney, but I hadn't been able to sleep for worrying about what was going to happen. I'd been a federal and state fugitive for seven and a half years, and I was going to turn myself in today.

I opened the *New York Times* to the front page. *Jesus Christ!* Staring out at me was a feature article, titled "Rudd, Who Led Campus Revolt, Will Surrender." Flushed, I speed-read through the article about my involvement in antiwar demonstrations at Columbia University in 1968, the rise of Weatherman, the bombings, the years underground, the decline of the movement. There didn't seem to be any indication that the prosecutor would renege on the deal my lawyer had arranged.

Going back to the beginning, my heart stopped as I took in

the interview with my father that I had skimmed past the first time. Jake got the Quote of the Day: "He's thirty years old. You get too old to be a revolutionary; it's time to start something new." I didn't know whether to laugh or cry. I had asked my parents not to say anything at all to the press, explaining to them that I was going to keep silent. I would not play out the role of repentant middle-class white boy who was now repudiating his past. I might have been wrong about a lot of things, but I'd been right in opposing the war and about the antiwar movement, which had played an important role in ending it.

On the other hand, I would not trumpet the disastrous strategy of armed revolutionary struggle, which had led me to founding the Weather Underground. How could I? By 1977 I saw the underground as a total failure as well as a tragic mistake. Nor did I still believe in an imminent socialist revolution in the United States, as I once had. I had no intention of resuming the role I'd played in 1968 and 1969, media symbol, spokesman, and leader of the student antiwar movement. It had been a long time since I could speak for anyone else, and I never did enjoy the posturing that came with the job.

Most important, by not talking I signaled to the government that I wouldn't give information against my former comrades who were still underground. Silence was my only possible strategy. And I reasoned that if I wasn't going to talk, my parents shouldn't either. Of course I underestimated them; keeping silent was not in their repertoire. Besides, why would Jake listen to me? From his point of view, I had totally screwed up things on my own. I understood exactly what he was thinking. He was trying to do for me what I refused to do for myself: repent in order to get off easier. He was hoping I would be welcomed back, my violent and revolutionary outburst ignored as a childish aberration. He was playing his role as The Father in the press's morality play of the prodigal son who returns, repents, and is forgiven. Downcast, I realized that it was going to happen whether I participated or not. And I would be forced into the role of a middle-class hero no matter what I did.

My father's apology would serve in place of my own. I got off easy.

Half a year before, in the spring of 1977, I had been anxiously watching the stairs at the end of an elevated subway line in the Bronx from inside a candy store. Trains came and left. Several casually dressed young men had gone up the stairs; their transistor radios with antennas catching my attention. What if those were walkie-talkies? I was worried. *What irony if I got busted in the process of arranging to turn myself in.*

Suddenly Gerry Lefcourt, my attorney, appeared. Still trim, wiry, and youthful, he hadn't changed much in the seven years since I'd seen him. I later found out that the former playground b-ball player had taken up marathon running.

I left the store and followed Gerry at a distance of a block, watching other pedestrians and cars. He turned a corner, just as he'd been told to do, and I waited to see if anyone followed. Everything seemed clear (but how could I really be sure?), so I sped up and caught him several blocks later. We hugged, delighted to see each other after so much time. I led Gerry to a nearby park where Sue was waiting with Paul, now almost three years old. We joked about the two subways and a bus Gerry had had to take to get to the meeting. Then we got down to business.

Sue and I filled him in about the years we had spent under, both closer and further from the organization, of my disillusionment with the leadership since the time of the town house, then my realization that the strategy of armed struggle had been a mistake. I told Gerry that I did not want the government or the media to use my coming in as a symbolic statement of the death of the movement. On the other hand, I didn't want to defend the Weather Underground, so I would say nothing.

Gerry said he thought that this was a good time to come up, especially because the new Carter administration was attempting to put the war in the past. Although it didn't directly pertain to my case, one indication of this friendly climate was the recent establishment of an amnesty program for draft dodgers and deserters. Carter's election indicated that the country wanted to forget the traumas of Vietnam and Watergate. The new president had brought in a new, less vindictive Justice Department. Internationally, Carter was trying to project a human-rights-oriented foreign policy as an alternative to the military force used by Nixon and Johnson before him.

Gerry believed that the remaining state charges were not too serious. In his very methodical way, he outlined four possible sources of legal difficulties: (1) federal warrants, which might not have been dropped back in 1973 when the charges were dismissed; (2) the 1969 Niagara Falls marijuana-possession charges; (3) 1968 trespass charges from Columbia; and (4) the Illinois state charges of riot, incitement to riot, and assaulting an officer from the Days of Rage in 1969. All of the various authorities in these cases would have to be contacted to negotiate an orderly surrender.

We agreed that advance publicity would be bad, because it might result in higher bail or denial of bail. We decided to keep the whole process secret until the end of the summer, when we felt we'd be ready. Sue and I did not tell Gerry that we were waiting because of the problem of JJ's security. Together, we set an arbitrary date of September 14, 1977, to turn myself in.

At that time we were living in an apartment in Germantown, a neighborhood of Philadelphia. Needing to flee New Rochelle the previous December, I had found this old railroad flat on the first floor of an attached duplex row house so typical of Philadelphia. There wasn't too much light, the windows giving out onto the walkway between houses, but the ceilings were high and the place was clean. The landlord, Jim, a genial African-American man in his late forties who worked as an accountant for the Defense Department, lived upstairs. He seemed pleased to have a young hippie family—mom, dad, and two-year-old—as tenants.

After putting a deposit and first month's rent on the apartment, we had barely any money left. We'd had to abandon our Chevy Nova, donating it to the street strippers back in the Bronx, so as to leave no tracks to our new home. Our most immediate need was *work.*

It was my turn to find a job, since Sue had worked at the day-care center in New Rochelle while I'd stayed home with Paul. After weeks of trudging the streets, looking for anything that didn't require a background or work history, I finally landed a job at a garment factory that manufactured knit polyester women's apparel, mostly shirts.

I was put onto a small crew tending a humongous cloth printing press. My co-workers were young guys, mostly high-school dropouts. We all made minimum wage, $2.55 per hour with a quarter raise after forty-five days. Later, probably because of the fact that I never missed work and could read and write, I was transferred to running the warehouse involved with delivering cloth to various locations in the factory.

On my rounds I would frequently stop and schmooze with people, asking questions about them and their work. Often we would get into long, involved conversations, sometimes on political or philosophical issues—the war, Watergate, fashion, the consumer society. I got to know the aristocratic, high-paid cutters, the black and Latina female sewing-machine operators, the guys in the shipping department, even the designers in their own little wing of the building, with their artistic pretensions. Though I was thrilled with my role of secret amateur sociologist in that factory, almost everyone who worked in the plant was unhappy, even the all-powerful managers, judging by how often they screamed at us and at each other. Whenever a dispute would arise, they would threaten workers by telling us the plant was in danger of being closed down and the work sent out overseas. This shut us up.

Twenty years later, happening to be in Philadelphia, I drove my family back to show them the factory I used to work in. I couldn't find it, though I knew the neighborhood well; I seemed to be lost. Finally I realized that we were on the right street and the right block but that there was an empty lot where the ancient redbrick six-story fortress had been. Management's threats had come true—they'd closed the factory, probably moving the work overseas. I wondered what had become of all the people I'd once known there, and their families.

That summer was one of preparation, mostly mental. My biggest worry was prison. I decided that I could handle up to two years' time if I had to, though I didn't know how I'd cope with more. Sue and I both wondered about the consequences of a long separation from Paul; he and I had never been apart in his first three years.

Toward the end of the summer, Sue started talking about having a second child. We had always told ourselves we wanted more than one, but I had assumed we'd wait until things settled down, after I'd surfaced. Now Sue said she wanted to try right away, just in case something happened to me. I was surprised by her insistence, but as usual I trusted her judgment. By early September we knew she was pregnant.

I was constantly anxious those last months in Philadelphia, worrying about what would become of us. Between the anxiety and my physical labor in the garment factory, I lost twenty-five pounds. I quit my job in August. We told our new friends and neighbors that we were moving to California, that old chestnut of a lie, hoping that this would be the last time we'd have to sneak out of a community. Right before Labor Day, we packed our few remaining possessions—a potted Chinese evergreen we had lugged around the country for seven years, a frying pan bought at a flea market in Santa Fe, Paul's clothes, toys, and books—in an old Falcon station wagon we'd bought and headed east, across the Delaware. We wanted at least a week to elapse between leaving Philadelphia and surfacing. We would pass the time camping at various state parks in New York and New Jersey.

From a phone booth out on the road, we contacted Gerry in a prearranged call. He sounded optimistic. The FBI and the Justice Department were in disarray, plus the Chicago and New York City cases had been put into "inactive" files. No one seemed particularly vindictive or gung ho to prosecute. Best of all, the D.A. in the Niagara Falls case said he felt that the charges had been trumped up. He told Gerry, "Don't call us, we'll call you."

We agreed that I would meet Gerry in his office at 8:00 A.M., September 14. I hung up and went back to the car to tell Sue, my heart beating with fear and anticipation.

The night before I was to turn myself in, Sue, Paul, and I were staying at the home of a friend in the Bronx. That day I had disposed of the Falcon by parking it on a street in upper Manhattan. Step by step we had destroyed all connections to the past, yet we were in a momentary limbo between lives. There was no going back, and what lay ahead was totally unknown.

I was in the bathroom shaving off the beard that had protected me the last seven years. As I watched the whiskers flush down the sink drain, I saw part of my identity leaving me. Tomorrow, barefaced, without my mask, I would return to being "Mark Rudd," whoever that was. Suddenly three-year-old Paul came running down the hall. "Daddy, Daddy, you're on TV!" he was shouting. "Come look!"

Sue and our friend were staring at the box, stunned. The local news was showing an old picture of me with a superimposed title reading, "Radical to Surrender." The anchorman said, "At ten-thirty tonight on Channel 4, we'll have a special thirty-minute report on the revolutionary career of Mark Rudd."

"So much for secrecy," Sue said. Someone had blown the story, and now the media attention might change the whole deal. I probably wouldn't be released as easily as we had hoped.

Unable to sleep, I was up before dawn and down at the subway stop by 6:00 A.M. I checked my new surfacing disguise one last time: My hair was short and my face clean-shaven; I was wearing a blue-checked dress shirt, chino slacks, and my old imitation suede jacket. I also had on clear eyeglasses so that any new pictures wouldn't be recognized by people I had known when I'd lived underground.

This was the moment I'd been anticipating for years. I bought a newspaper, the train came into the station, and I got on.

At nine o'clock exactly, Gerry and I walked up Centre Street toward the Tombs. A block ahead we saw a loud jostling mob of at least several hundred people; we thought some kind of demonstration was taking place at the courthouse. As we came closer, men with press badges and cameras ran toward us. "Mark, will you give us an interview?" screamed the first reporter as he thrust a microphone into my face. "How does it feel coming back after all these years?" *Whoa, this crowd is here for me!*

Dozens of people were now pressing in on us, elbowing one another. In the middle, I seemed to be pushed along like a stick in a current, unable to respond in any way other than, "Excuse me, excuse me." At one point Bob Feldman, my old friend from Columbia SDS, appeared in the jumble, and we grabbed each other's hands; I was touched that he was there. All the while, more questions were being hurled at me. "Mark, why did you decide to turn yourself in?" "Mark, what have you been doing these seven years?" "Mark, give us some of that old rabble-rousing style!"

"Nice weather," I replied as I reached for my back pocket to make sure that no one had boosted my wallet.

Somehow a path opened in the crowd, and Gerry and I stumbled through to the side door of the courthouse. Remaining more or less the calm in the eye of the storm, I had come through unscathed, but somebody had elbowed Gerry right in the ribs. "That was worse than any riot at Columbia," he quipped to me.

I had spent many hours, even whole days, in the lobby and the cells of the Tombs, beginning with my first bust at an antiwar demonstration in 1967. Now, ten years later, the smell, the feel of the place, was all too familiar: the ultra-worn stone floors, the old wooden benches permeated with the sweat of thousands of bodies compelled to be here because they or their loved ones were in trouble with the law. Here I was, returning to the scene of old "crimes," back again in the government's clutches after years of freedom. I was scared.

Roughly five hours later, I was standing in front of Judge Milton L. Williams for my arraignment. Up to now nothing had gone wrong. In fact, I was surprised that I was being treated more or less as a VIP, rather than a bail jumper and an accused felon revolutionary. I had the distinct feeling that people I encountered—assistant D.A.'s, clerks, guards—looked at me with curiosity, as if I were a celebrity. I waited several hours in the district attorney's office until my fingerprints came back from Washington. No one attempted to question me, nor was I shown any hostility.

When the fingerprint check finally came back, proving conclusively that I really was Mark Rudd, guards handcuffed me and led me away. The holding cells were crowded with black and Latino men, all waiting to appear before a judge. I was put into the only empty cell, where I sat alone, waiting and wondering what would happen.

After a few more hours, a guard came and got me. As I walked down a corridor to the courtroom, past other cells, Gerry came through the door. "After seven years it's still WSP [White Skin Privilege]," I whispered to him, motioning toward the cells filled with nonwhite men.

You are never more helpless than when standing in front of a judge: the whole setup is designed to tell you that all power lies with this man and the state behind him. Still, I felt I was ready for whatever would be given me. The D.A.'s office had agreed that I would be released on my own recognizance. But all might not go as planned.

Judge Williams, a light-skinned black man, addressed Gerry and the assistant D.A. handling my case, a woman. "My concern is that be he rich or poor, black or white, a small person or a large person, no bail jumper should be lightly allowed to walk out of my court," lectured Judge Williams.

The whole process that day had been defined by privilege—my being able to come back as the returning middle-class white boy handled with kid gloves. A judge was considering whether to let me go free, without even posting bail. Perhaps he was thinking of the holding cells just outside the courtroom filled with black and Latino men who could not make bail; I know I was.

The assistant D.A. was talking. "Judge, the preparations for Mr. Rudd's surrender have been very elaborate, and the district attorney's office is convinced that he intends to see this case through." This D.A.—whom I would have called a "pig" a few years before—was arguing for me.

The judge capitulated. "Well, if both sides agree to no bail, I'll have to give in. Released on his own recognizance!"

I hadn't said a word the whole time. I didn't need to. Everyone in the courtroom who bothered to pay attention knew that the fix was in.

The guards let Gerry and me out a back door of the courthouse in order to avoid the swarm of reporters in the lobby. Gerry's secretary was waiting on the street behind the Tombs in his BMW. She drove me a few blocks to the World Trade Center, where I caught a PATH train to Newark, then a city bus up South Orange Avenue. I called my folks' house, and my brother, David, came and picked me up at the bus. I hadn't seen him in more than seven years. We looked at each other, amazed.

He drove directly into my parents' garage around the side of the house, because the front door was besieged by reporters and cameramen. It was 6:00 P.M., and I was home again, just in time for dinner.

There was an air of pandemonium and hysteria in the house. Bertha and Jake were cooking, as usual, but outside at the front door reporters and photographers were ringing the bell to see if my folks would come out and talk. They had not seen me come in. The phone rang constantly with well-wishes from relatives and my parents' friends from their synagogue. But most of my parents' hysteria was aimed toward me. Fear, anger, and love were all intimately tied up in this homecoming. David later explained their reaction to me. "It's like seeing your child almost hit by a car. You run to embrace the kid, but instead you grab him, screaming, and shake him."

They both cried, my mother telling me what she'd probably wanted to say for years: how much they'd suffered throughout the time I'd been in hiding. I cried, too, and told them, lamely, that I was sorry for all the pain I'd put them through, even though I thought what I was doing was right. "I hope it's over now," I said.

That night the entire Rudd family sat at the round kitchen table, and while we talked of hurt and guilt and the war, the chicken soup and matzo balls appeared, plus a steady stream of Jewish appetizers—pickles, chopped liver, coleslaw. (A few days later, a reporter for the *New York Post* got through to my mother and obtained her recipe for chicken soup, including her secret ingredient—a whole sweet potato.)

"It's not as if I haven't eaten in seven and a half years," I told my mother. But that was beside the point. I hadn't eaten *her* food in that time. Such a thing was incomprehensible in my family, like going to the moon, or divorce, or marrying a gentile. Bertha had visited or called her mother every day of her life. She expected to be treated more or less the same way by her kids. I had violated that basic covenant by leaving for so long, and under such harrowing conditions. My mother told me she had cried every night at ten when the nightly news asked, "It's ten o'clock. Do you know where your children are?"

I tried to lighten the mood with the happy news that Sue was pregnant with our second child. Jake responded with an offer he'd been thinking about perhaps his whole life—that I would join his business, apartment investment and management, and eventually take over from him. Having worked from the time he was a child, he was almost ready to retire now at the age of sixty-seven. The offer wasn't merely a job. It was an opportunity for me to redeem myself and, at the same time, a total vindication of my parents' life choices. It would be their perfect ending to their long nightmare.

For my part, accepting would mean total surrender beyond the fact of turning myself in, a confirmation that I had been deluded in my desire for revolution. I could become my father. Just a few weeks before, I'd been one of the underclass, a minimum-wage laborer in a factory owned by Jews. Now I could take my place as one of the owners, worrying about the *schvartzes* ruining my property values.

I needed to say something. I told them I might continue my education or I might get a job in construction, which is the trade I was best suited for. "But I don't want to be in the business," I said with finality.

My father replied that I could do construction for him.

I said, "No, I don't want to be the boss's son."

"You're a schmuck," my father said contemptuously.

My mother offered her old refrain. "You could go to law school and defend people, like Gerry Lefcourt."

The next day I flew to Chicago to appear in court on my outstanding charges there from the Days of Rage. Again I was released on my own recognizance, with the complete support of the Chicago state's attorney. It was a far cry from the time in December 1969 when I was jailed for two days for violating bail restrictions by traveling out of Chicago.

I returned to New Jersey and stayed a few more days with my parents before reuniting with Sue and Paul. They were staying at the home of old friends in Brooklyn. We had temporarily separated so that the press wouldn't find out about my family—and they wouldn't become targets for some demented right-winger.

Sue and I moved into an apartment in Park Slope, Brooklyn, where many old New Lefties lived. We wanted to give New York a try. It was, after all, where we had started out ten years before, and many old friends from Columbia and SDS were still there. New York City was also close enough to my folks that we could spend time with them and they could get to know their grandson. Also, we needed a safe and comfortable place to have the new baby. We figured we would spend at least one year here, and if nothing worked out for us, we could move somewhere else.

The first month I went to visit old friends and political associates in order to catch up on developments of the last years and, more specifically, to see if there was a place for me in the movement. The seventies had been a time of contraction. Where once there had been huge antiwar organizations, by 1977 there was virtually nothing. My old friends carried on in tiny groupings, though much of the work was by now rote and automatic, rather than dynamic. They were keeping the faith.

I met Liz Fink, a dedicated attorney who had been working on the aftermath of the 1971 Attica rebellion, defending the Attica Brothers and fighting for the legal and civil rights of prisoners. One day, dropping in for a visit at Liz's office, I met Frank Smith, known in prison as "Big Black." Black had been a leader at Attica and was finally now out on parole. He was Liz's associate in her office, and we got to talking.

"You know," he said, "your situation is a lot like that of a con who just got out of prison. You don't really know what life is like in the outside world. You have to take time to look around, see what you've missed, how you've changed, and how you fit in. Relax, don't push it."

Black had picked up on my anxiety and impatience.

"I never thought about myself that way," I said. "It makes sense."

So I started out in Brooklyn by doing what I knew best, work. I got a job with a small contractor in Park Slope who remodeled brownstones for the young professionals who were just then beginning to flood the neighborhood. I framed and drywalled and painted and hauled trash from the opulent new homes that had been rooming houses for the poor. Very quickly I grew contemptuous of the materialism of our customers: their designer kitchens with granite floors and countertops, restored oak woodwork, backyard cedar decks. Sue and I began referring to these lawyers, professors, and businessmen as "the hipoisie." Later the press would anoint them forever as "the yuppies."

I worked, but my heart wasn't in it. I grew more and more depressed, worse even than my dark times when I first became a fugitive. One day I actually found myself thinking how comforting it would be to throw myself in front of a subway train. I told Sue how bad off I was, and she said that if I ever committed suicide, she'd never forgive me. That did it: I didn't want to encounter Sue's eternal wrath.

A friend laughed me off when I told him I was suicidal. "Jewish princes don't commit suicide," he quipped. "At worst they might eat themselves to death."

Part of the problem was that I was living a lie. I seemed to be continuing my personal myth of "proletarianization," as if I had actually become a working-class person in the years underground. But the truth was, I still had family resources, money available, a partially completed education, and the potential to do almost anything I wanted. Most important, I had returned aboveground to become an activist again. Yet here I was sanding hardwood floors, not affecting society. It wasn't as if the United States had suddenly become a paradise on April 30, 1975, when the Vietnam War ended.

I was drifting. Old comrades were friendly enough, yet I couldn't get excited about their work. For example, solidarity with Puerto Rico and Chile, both victims of U.S. colonialism and intervention, was important, but after all the difficult work of meetings, setting up forums, and demonstrations, it affected so few. It seemed as if these leftists of the late seventies were existential warriors, continuing to fight for the sake of their own souls. I respected them—still do—but had not yet found my way.

A few people closer to old Weatherman politics would have nothing to do with me. Once, on a bus going to work, I found myself sitting across the aisle from a comrade from the Weather collective and first underground days, back in late 1969 and early 1970. This woman and I had once been lovers; I had adored her. I'd heard she was now with the May 19 Communist Organization, named for the day of both Ho Chi Minh's and Malcolm X's birthdays, the ultra-correct successor organization to the Prairie Fire Organizing Committee. In other words, she still actively supported the fundamentalist revolutionary armed struggle and solidarity line I had rejected by coming up.

I tentatively said hello to her in a quiet voice. Agitated, my old lover nodded, got up from her seat, and moved to the rear door of the bus, jumping off at the next stop. Perhaps she was scared of catching a dangerous right-wing infection.

One New York City group where I was warmly welcomed, however, was the War Resisters League (WRL), the ancient pacifist organization that was Dave Dellinger's base. These people had unique politics—opposing all wars and violence but at the same time supporting liberation movements here and around the world. They felt that they had no right to dictate to people how they should fight for their freedom. On the other hand, they always advocated nonviolence. Hanging out at the WRL office on Lafayette Street in Manhattan, I met old-time socialist-pacifist radicals like Dave Dellinger, David McReynolds, and Ralph DiGia, people who had been in the struggle for decades before I came on the scene. They treated me with human decency and seemed to understand the conflicts I had lived with during all the years underground. I found that I could talk with them about both my history and their nonviolent ideas. In time I would join WRL and take their pledge "not to support any kind of war."

I easily settled both my criminal cases. In New York I copped a plea to a misdemeanor charge of criminal trespass stemming from the occupation of one of the buildings at Columbia in May 1968. In return, all the other charges were dropped. My sentence was unconditional discharge, the same as everyone else received from that incident.

The Chicago charges were also reduced to misdemeanors, and I pleaded guilty to two counts of aggravated battery, in return for which the other charges were dismissed. Even though I was the one jumped by the cops, the deal was so good that I figured I had nothing to argue about. I got two years' probation and a two-thousand-dollar fine.

I was allowed to serve my probation in Brooklyn, where I would report once a month to an elderly probation officer in a downtown office. I usually took Paul with me, and the boy so charmed the old man that soon the visits became more like a grandson visiting with his grandfather. The other probationers waiting to see their POs, black and Latino men for the most part, did not bring their kids with them.

By the time our daughter, Elena, was born in May 1978, it had become obvious to both Sue and me that we had little going for us in New York City. Living there was very difficult, compared with the places we had lived underground. It only makes sense to endure the hardships of the city if there's some overriding reason to be there. Life was especially hard with little children, who had to be taken out to the park, watched every minute so they wouldn't play in dog shit or broken glass, and protected from the dangers of the decaying city.

Though many people in Park Slope befriended us, we still didn't feel as if we were part of a community, which was one of our goals in coming up. In fact, we wondered whether such communities were even possible in a city where people struggled so hard to survive. Most of our friends were stuck on the subway, job, subway, then home-to-collapse treadmill and had no time just to hang out, which is how we were used to living. And group, communal activities were almost out of the question; you had to make plans weeks or months in advance.

Since leaving New Mexico years before, Sue and I had kept imagining ourselves back there. We missed the people, the land, the open sky, the chili. We also had an idea that in a place with a small population, radical political organizing could have more of an impact. By the summer of 1978, we decided to move back. For the absolute last time, we hoped, we would pack up and go—only this time with two kids, a four-year-old and a newborn.

The hardest part was leaving my parents again. I loved them dearly for the love that they gave me, also for the mixed messages, the hurts, and the anger. But I had to go west again, the traditional way Americans start a new life. They stood out in front, tears in their eyes, as we pulled away from their house in an old Plymouth station wagon with a bum transmission that my father had bought us for six hundred dollars, dragging a U-Haul trailer behind.

Epilogue

When Sue and I arrived in Albuquerque in August 1978, the first thing we did was join the local antinuclear movement—opposing nuclear waste dumping and uranium mining in New Mexico. Our pent-up energy for organizing and community had finally found a channel for release. Just as we'd hoped, our little rented house next to a giant cottonwood tree became a center for strategy meetings, informal discussions, and impromptu parties.

I enrolled at the University of New Mexico in a teaching program—after convincing a wary admissions officer that my intention wasn't to stir up trouble but to get a degree so I could support my family. By June 1980 I had earned a bachelor's degree and immediately began teaching English and reading at the local technical-vocational institute. My construction experience in the years underground helped me get hired.

Meanwhile, our marriage was going through rocky times. Before we moved, I had made an exploratory trip to Albuquerque. After a few days, I called Sue back in Brooklyn to report. "It looks like there's a lot for us here," I said. "Good people, the university, lots of work. Only if we move, we'll almost certainly get divorced. There seems to be something in the air."

"I don't care," she answered, laughing. "I want to live in New Mexico."

Sure enough, within a year of arriving in New Mexico, both of us were talking of separation. It took another year for the Sturm und Drang of the breakup to play out. In October 1980 I moved out of our rented house, into a little one-bedroom apartment downtown.

For over ten years, Sue and I had been each other's best and only friend. We had faced danger together, solved all sorts of dire problems, experienced the loneliness and isolation of life as fugitives. Sue had never regretted her decision to join me in the underground, but the ordeal had taken its toll. Now that the external constraints were off, we needed to get away from each other and start new lives; we craved relationships with other people.

For my part, I knew I owed my life to Sue. Had I been forced to face being a fugitive alone, I most likely would have fallen into depression and even suicide. But I had Sue to love, to laugh with, to build a family with, to keep me sane.

Yet I was haunted by memories from the years underground, especially old hurts, resentments, guilt. Only after the separation did I realize that I had needed to break with the past, escape the catastrophe of the Weather years, and go on to an entirely new life. I suspect that Sue's feelings were similar to mine.

Paul was six when I moved out, and our daughter, Elena, was only two. The first thing I did in my little apartment was build a bunk bed for them. We all slept in the same room, which comforted me. Sue and I amicably worked out a joint-custody arrangement for the kids; they went back and forth between our houses daily. The advantage of our arrangement, from my point of view, was that the kids and I were never separated from each other for too long.

Sue and I have successfully co-parented our kids since 1980. We have remained friends, even as our separate lives took a variety of twists and turns. In 1984 Sue remarried; two years later she had a third child. Our goofy extended and blended family has survived well.

Teaching adults at the Albuquerque Technical-Vocational Institute (TVI)—it eventually became a community college—gave me a chance to address the underlying question of how people learn, a question to which I had been oblivious, unfortunately, when I joined SDS thirteen years earlier. I'd naively thought that "revolutionary consciousness" would grow automatically as millions of people became aware of the destructive nature of the system that gave us war and inequality. It didn't happen that way.

At TVI, I taught reading, writing, algebra, basic arithmetic, and problem-solving and communication skills for construction trades in a remedial or "developmental" program. All the students were high-school graduates who came to the institute for training in a vocational area, yet they lacked basic skills. Waking up every morning wondering what I would learn that day, I felt like the most privileged person in the world. I came into intimate contact with up to 125 students per semester, three semesters per year; in the classroom they opened up and revealed their fears and self-doubts. I had to figure out, often on a case-by-case basis, how to help my students learn.

If I had any political preconceptions left over from my ideological days—about workers, women, Chicanos, Indians, or any other social category—teaching at TVI cured me. Over the years I marveled at the never-ending parade: Vietnam vets, Vietnamese refugees, junkies, dope dealers, cowboys, teenage punks, single mothers, displaced homemakers, active-duty soldiers, retired military lifers, alcoholics, people with back injuries or neurological disorders, paraplegics, the learning-disabled, born-again Christians, heavy metalists, ex-cons, the homeless, abused women and the guys who beat them, neurotics, psychotics, ballet dancers, sculptors, artists of all varieties, geniuses, morons. They taught me endless lessons.

In the late eighties, I brought together a group of teachers at my school to reform our math curriculum and pedagogy, which was quite traditional. Part of a nationwide movement, the general idea was to substitute actual thinking and real-life problem solving for the rote learning of math procedures. Also, starting in 1990, I helped organize a union for faculty and other employees. After a long and difficult struggle with the administration, we won our first contract in 1995. I've stayed active with the union since, primarily working in our local's Committee on Political Education, which does electoral work.

A year after Sue and I separated, in October 1981, I got a call from my father. He was hysterical. He had just heard on the radio that Kathy Boudin, my old friend and Weather comrade, had been busted in an attempted armored-car robbery in Rockland County, upstate New York. A Brink's guard and two cops were killed. "The blood of those policemen is on your hands!" he screamed at me across two thousand miles.

"What are you talking about?" I asked.

"It's your stupid ideas that started all this, and it's my fault I let you get away with it!"

"It's not likely you could have stopped me," I said, more or less rationally. But I was trembling inside. "Look, you're not responsible for anything. Maybe I am, I don't know. Let me get today's paper."

Arrested along with Kathy was another friend, Judy Clark, from the former University of Chicago SDS chapter, who had also been a founder of Weatherman. The front page of the *New York Times* ran a picture of a dazed Kathy, gaunt and with long, curly, dark hair, being led into a police station. Beneath was a smaller, fuzzier shot of Judy Clark, and there was a third picture of an "unidentified suspect," a male with a prominent, bashed-in nose and a very dark, full beard and thick head of hair. Both Judy and the man had black eyes and looked as if they'd been badly battered. I stared at the man, trying to recognize him. Slowly it dawned on me that this was David Gilbert, my old mentor from Columbia.

According to newspaper accounts, the armed robbers, all black, stopped the armored car outside a bank in a suburban shopping mall. They shot one of the three Brink's guards when he went to draw his weapon. The robbers then escaped in a red van with sacks of cash totaling $1.6 million. Minutes later, tipped off by a call from an eyewitness, police stopped a rented U-Haul driven by David Gilbert at a roadblock at the entrance to the New York State Thruway. Beside him in the passenger seat was Kathy Boudin. Guns drawn, police surrounded the vehicle and forced the two white people out of the front. Suddenly the back doors opened, and an undetermined number of black men jumped out, automatic weapons blazing. In the ensuing firefight, two Nyack village police officers, one white and the other the only black man on the force, were killed. No one but Kathy Boudin, unarmed, was arrested at the roadblock.

Two other getaway vehicles managed to flee the shoot-out, but one, driven by Judy Clark, crashed into a stone wall a few miles away after a high-speed chase, and David Gilbert and Sam Brown, a black man, were arrested along with Judy. The other car actually escaped.

Leads uncovered in the bust disclosed a substantial network of apartments and cars in New York and New Jersey. Apparently my old friends had hooked up with the Black Liberation Army, an ultraradical group descended from the New York chapter of the Black Panthers. Numerous associates and supporters of the gang were arrested or became fugitives, including members of the May 19 Communist Organization, a white group descended from the Prairie Fire Organizing Committee—a front for the WUO—of the mid-seventies. A few days later, there was a shoot-out in Queens, New York, in which one member of the group was killed and another arrested.

I was stunned: My friends seemed to be involved in a very different league from the old Weather Underground. Kathy and David had a fourteen-month-old infant, named Chesa, whom they had left at a babysitter's the morning of the robbery. Eventually the child went into the custody of Bill Ayers and Bernardine Dohrn, who raised him in Chicago as their third son. Their two other children had been born underground before Bernardine and Billy surrendered in 1980. Raising Chesa was difficult; he was an abandoned child with terrible emotional problems. But now he's a grown man, a brilliant writer, organizer, and speaker on U.S. imperialism in Latin America. Chesa advocates for children growing up with incarcerated parents, among other issues. He knows what it is like to be strip-searched as a five-year-old or to be forbidden by a guard to touch your father during a visit.

Gathering evidence for the Brink's trial, a 1982 grand jury called Bernardine Dohrn as a witness. Bernardine understood the proceedings to be a coercive fishing expedition and refused to testify on the grounds that the grand jury was unconstitutional. She was jailed for contempt of court for seven hard months that kept her away from her three little kids, all under the age of three.

Under the "felony murder" law, there is no distinction between those who pulled the triggers and those who were unarmed accomplices. At their murder trial, David Gilbert, Judy Clark, and Kuwasi Balagoon claimed political-prisoner status under international law and refused to participate in their trial. All received sentences of seventy-five years to life, with no possibility of parole. Kathy Boudin chose to plead guilty to one count of felony murder and robbery in return for the possibility of parole after twenty-five years. Word had it that she was a last-minute, ambivalent recruit to the holdup plan. At her sentencing, Kathy expressed her deepest regret at the loss of the lives of Brink's guard Peter Paige, and the two police officers, Waverly Brown and Sergeant Edward O'Grady, Jr. Between them they left nine children without fathers.

What we started back in 1969 and 1970 had led, step by step, to this tragedy more than a decade later. There did not appear to be much difference between what Judy and Dave proclaimed about the robbery and what we might have said twelve years before. They claimed that this was a "revolutionary appropriation" intended to aid the liberation of the black nation in America, that the arrested were prisoners of war and political prisoners who should not be tried in any U.S. court, that the conditions of oppression of black people and of colonized people throughout the world made the U.S. government criminal and necessitated armed action. At one point during her trial, Judy Clark modestly stated that the Brink's robbery was the most important act in U.S. history since John Brown's raid on Harpers Ferry. My friends held the same exact analysis and advocated the same strategy—armed struggle—as we had in 1970, with even more disastrous results.

After his trial and conviction, in the spring of 1981, David Gilbert called me collect from prison in upstate New York. He was friendly, curious about my family and work, as sweet and caring as he had always been. But I couldn't contain myself. After only a few seconds, I began screaming at him, across years and prison walls. "What did you think you were doing?" I heard myself say. "Did you still think there's a revolution happening in this country? Black people aren't revolutionary anymore, nobody is. Can't you see reality?"

"You don't deny that the U.S. is still imperialist, do you?" he calmly asked.

"No."

"You don't deny that this is still a racist country, do you?"

"No."

"Well, someone had to keep the revolutionary underground going. We couldn't just surrender," he concluded the argument.

I was lost. "But it's not real!" I screamed. "What revolution? You're totally isolated, out of touch. You were underground so long that you have no idea what real people think."

"Oh, yeah? When you're in prison, you get to see who benefits from this society and who's hurting. Look, I didn't call you to argue," David said, quite reasonably.

I was shaking. "I'm sorry," I said. "I didn't intend to pick a fight."

Begging off because I was late to work, I promised to write him. It took me thirty- four years to figure out what to say. My emotions were too raw. Dave had been my twin self in 1966, my moral hero who didn't want to be a good German; from 1969 on, he was the self-sacrificing revolutionary guerrilla, forever loyal to a heroic idea of using cleansing, pure, logical, revolutionary violence to stop the greater violence of the system. I, on the other hand, compromise every day of my comfortable, safe life. A Columbia graduate and brilliant sociology graduate student, David might have had a career as a respected professor, talking a great game about revolution. But he believed in throwing in his lot with the oppressed of the world, no matter the consequences.

In prison David has been remarkably productive over the years. He founded an AIDS peer-counseling program that has been credited with saving hundreds of lives, and for years he also tutored prisoners who wished to earn their GEDs. He continues to insist on the original revolutionary justification for the robbery, though lately he has been publicly expressing his personal regret for the loss of lives.

In 2007, after much urging by mutual friends, I overcame my emotional blocks and wrote David. We have been exchanging letters since then. He claims he does not remember the 1983 phone call in which I unloaded on him. He is still the same concerned, affectionate, brilliant man he always was. Photos of him in prison show him still looking uncannily young and hopeful, despite the harsh imprisonment.

Kathy Boudin, who had plea-bargained, was imprisoned at the Women's Correctional Facility in Bedford Hills, New York, Westchester County, for twenty-two years, between 1981 and 2003, when she was released on parole. In those years Kathy earned a master's degree in education, started a nationally recognized nursery for children visiting their incarcerated mothers, and taught parenting and AIDS-education classes. Many times over the twenty-two years, Kathy, acknowledging her own role in the tragedy, has expressed her regret for the deaths of the three Brink's victims and her apologies to their families. Judy Clark remains in prison at Bedford Hills, where she became Kathy's best friend, confidante, and coworker. Based on a newfound Jewish identity, she underwent a profound spiritual reorientation and consistently expresses regret for her participation and advocates nonviolence.

In the many years since the Brink's robbery, I've kept repeating the question over and over in my mind, *What's the difference between David and Kathy and Judy and myself?* Rationally, I knew that I had gotten out of the underground in 1977 and that I had left the ideological cult of armed struggle even earlier. Yet, more deeply, I felt that there was no difference. I loved these people, they were my intimates, and what we'd started together they had merely continued straight to the tragic end. My father was right: I did bear some responsibility.

The violence of the era—in Vietnam and on the streets of this country—and the tragic consequences have to be looked at clearly and rationally. We need a truth and reconciliation process, such as happened in South Africa, to finally put an end to the suffering.

In 1982 I began building a house with my kids. I wanted a project for us to work on, something more than just going shopping at the Safeway. A friend had a defunct six-acre farm in the South Valley of Albuquerque, a semirural Chicano barrio not far from downtown. He offered to sell me the old half-acre orchard for almost nothing. I immediately started clearing away the weeds and debris.

The house was to be a passive-solar adobe, modeled on both traditional New Mexican building and solar designs of the late seventies. I had never built a whole house before, but I had enough construction experience from my years underground that I knew what questions to ask. I've always had a knack for getting people to volunteer. To dig the trenches for the foundation, I threw a party on Memorial Day and told the guests to bring picks and shovels. The digging got done in a day, and everyone had a blast. The next month, during a giant two-day party, we made two thousand adobe bricks, almost enough for the entire house. Dozens of kids jumped into the dirt pit and mixed the adobe mud with their feet.

Over the next two years, I kept up a steady pace of building on weekends and vacations, taking the winters off to recharge my batteries. I found an entire community of owner-builders in the South Valley, mostly Chicano guys who helped each other through giving advice or trading labor. One day, after about two years of work, I was complaining to Simon Ortiz, a poet friend from Acoma Pueblo, that the house was taking forever and that I wondered if we'd ever move in.

"Two years is nothing," he said, laughing. "There are houses in my village that have been under construction for fifteen or twenty years. Houses *should* take a long time to build."

He was right. By going slowly I was able to ponder each detail, making changes as I went along, according to the feel of the place. The house would not have turned out as well had I gotten a construction loan and paid a contractor to build it in three months. Time, I learned, is the missing element in conventional building.

The kids and I finally moved into the house in the summer of 1984, even before the interior details were finished and the exterior was plastered. It's been a continual work in progress ever since. In 1997 I met Marla Painter, a community and political organizer from Nevada and California, and—immediately recognizing each other's hearts—we decided to marry. Marla needed a few changes in the house, so we moved out, gutted it, and rebuilt it to her specifications, this time taking on a mortgage. It's even a better house now, having been built twice.

Through the years I have continued to be involved in antinuclear, Native American solidarity, and disarmament work. When I realized, in 1985, ten years after the end of the Vietnam War, that the United States was attacking Nicaragua, I became profoundly depressed. To deal with my own form of posttraumatic stress disorder, I joined the Central America solidarity movement and devoted the next five years of my life to opposing the intervention. I helped organize the New Mexico Construction Brigade to Nicaragua, which built houses in that embattled country and then returned home to combat the lies of the Reagan administration about the Sandinista revolution. In February 1986, I was a member of the first of several brigades from New Mexico, helping to build four houses on a cooperative farm in the northern region of Nicaragua that had been attacked by the contras, surrogate forces clandestinely organized and paid by the United States.

Because the government did not use American soldiers in Central America, the American people did not have to pay attention to the war there, and by 1990 most of the revolutionary movements had been either stalemated or defeated, as in Nicaragua. Tired of years of war and privation, the Nicaraguan people sued for peace by voting in a pro-American government. Two decades of corruption and poverty have ensued.

I experienced another bout of profound depression in 1991, when the United States went to war over Kuwait. The antiwar movement—despite our protests—had little effect. Twelve years later, in the spring of 2003, when millions spontaneously poured into the streets to try to stop the second Bush's plans to attack Iraq, I was one of them. Bush and the neoconservatives in power merely ignored the protests, because the mass media and even the Democrats, allegedly the opposition party, didn't have the guts to vote against their lies. Peace activists once again began organizing for a long war. In Albuquerque I joined with other teachers to successfully defend the right to teach critical thinking and discuss the war in our classrooms. I continue to write and speak on the history of the Vietnam antiwar movement; my message is that we once actually did build a movement that helped stop another war of aggression—and that such a thing could happen again.

I also point out that the antiwar movement inherited from the civil-rights and labor movements a highly effective model for organizing—direct engagement between people. That was how we organized at

Columbia University from 1965 to 1968. Yet I also offer my own disastrous experience with violence in Weatherman and the Weather Underground, explaining how it didn't work and why it can never work in this country. Occasionally a young activist here or there catches on to what I have to say.

Fortunately, I didn't get depressed at the start of this war, nor did I in 2004 when Bush was reelected. I've accepted that lessons about the limits of American power are learned and then forgotten and then must be painfully relearned again. The current war is another opportunity to organize a peace and justice movement, as well a chance for a new generation of Americans to learn once again about the futility of wars of occupation and conquest. Perhaps it was my years as an algebra teacher—in which I repeated myself thousands of times—that led me to appreciate repetition as a necessity and a virtue. It's a fortuitous lesson, because organizing requires patience above all.

Since 2003, when the *Weather Underground* documentary came out, I've been talking with young student activists around the country about the differences between the Vietnam antiwar movement and the present antiwar movement. These young people are motivated by precisely the same feelings I was motivated by—grief over who our country is in the world and what we have become to ourselves. They often ask me, "Why isn't the current student movement larger?"

The culture of this country in the first decade of the twenty-first century is fundamentally different compared to that of forty years ago. There's no draft, so young people don't need to pay attention to the war, as we had to. (Which explains why a draft is not on the war planners' agenda.) The seductions of the entertainment and consumer cultures, fueled by cheap goods and easy credit, have achieved almost total hegemony as the purpose of individual life. To top this off, the cost of a college education, even at public institutions, is so much higher than it was forty years ago, when no one graduated with anything like the level of debt that students have now. Huge student loans keep young people so shackled that it never even occurs to them to stray from the career path for a protest or a meeting.

But the single biggest difference I find in talking with young people is what seems to be a pervasive belief that "nothing we do can make a difference." In the sixties I never once heard anyone utter anything like that. The example of the civil-rights movement in the South, which hit the media from 1954 to 1965, when so many of us were children, proved the opposite: What an individual does, in concert with others, *can* change the world. My generation was blessed with this knowledge.

Even with these enormous cultural impediments, a few young people have been waking up from the big sleep. The events of 2005 and 2006—the near destruction of New Orleans, the recognition by a majority of Americans that the war in Iraq could not be "won," the increasing awareness of the looming ecological disaster as a result of global warming—seem to have shaken quite a few kids who were in high school or even middle school. Like us in 1968, they realized that the people in power can't possibly be trusted to solve the world's problems. These young people have been entering college looking for student activism. A few have joined a new SDS—organized in 2006, with over two hundred chapters—or other student radical organizations. Even larger numbers formed the youth base of Barack Obama's presidential campaign of hope and change.

Most of these young people I meet call themselves activists. I almost never heard that term in the sixties, except when used by the media or a university administrator to refer to us as "mindless activists." We always called ourselves organizers. I was lucky to have fallen in with people at Columbia, like David Gilbert and Ted Gold among many others, who came out of the civil-rights and labor-union organizing traditions. The goal was always to build the movement, to "build the base." The main method was personal, one-to-one engagement between people.

It surprises many young people I talk with that the events of April and May 1968 at Columbia were the product of more than six years of concerted, focused, and unrelenting organizing, going all the way back to the Columbia chapter of the Congress of Racial Equality in 1962. I joined a gang that was in it for the long haul. Years of hard work by dozens of committed organizers at Columbia helped produce the changes in thousands of individuals' minds necessary for them to risk their college careers and take action. People don't join movements spontaneously; they join because they know and trust the organizers and because they've thought long and hard about their own personal relationship to the issues and to power.

After the success of the Columbia rebellion, many of my comrades and I got confused about why we attracted so much support. We thought that the breakthrough at Columbia was due to the Action Faction's in-your-face militancy, our willingness to take risk in confrontation with the university administration and the police. Unfortunately, we discounted the years of tireless organizing that had laid the groundwork.

We also didn't realize how lucky our timing was. The spring of 1968 blasted the country—one-two-three—with the Tet Offensive in Vietnam, the abdication of LBJ, and the assassination of Martin Luther King. Suddenly it became a moral imperative for thousands to follow the lead of black students in fighting the university's racism and of SDS in opposing its participation in the war.

Part of the tragedy of Weather was the irony that two of the leading theoreticians and proponents of organizing as base building, David Gilbert and Ted Gold, were seduced by the apparent success of militancy, probably for reasons not unlike my own, and abandoned their original insistence on the organizing model. Two years later Ted died in the explosion on West Eleventh Street; David Gilbert has been incarcerated since 1981 and—barring a miracle or a truth and reconciliation process—will end his days in prison.

The organizing model is also powerful because it recognizes that coalitions between dissimilar people are needed to achieve any sort of power. No small group can win a struggle alone. The civil-rights movement in the South was a grand coalition of black church people, students, unionized workers, white liberals from the North, and even, eventually, national politicians. Many young organizers I meet have never heard of coalition building, just as they've never heard of base building. Almost all accounts of the Columbia strike have omitted the fact that the rebellion would not have occurred without the coalition with the black students and Harlem. Their erasure from history—with the emphasis put on the white SDS—can only be ascribed to racism.

Slowly, however, a new generation of young organizers has begun to emerge. The media haven't yet discovered them, and when they do, it'll be a giant surprise, just as we were forty years ago. Having begun this dialogue with so many intelligent and committed young people, I suddenly realized in late 2006 that I was just going through the motions of my full-time job as a math teacher; all my mental energy was directed toward the problem of how to communicate this history. I felt terrible about shortchanging my students and my fellow teachers at Central New Mexico Community College. So I signed my retirement papers and began writing this book, all the while continuing the discussion with young organizers. There's no shortage of organizing work to be done.

On the weekend of April 24 to 27, 2008, exactly forty years after we closed down Columbia University, approximately four hundred people—mainly veterans of the occupation of the five buildings—came together for a reunion and commemoration of the event. We were all now around sixty years of age, give or take, and, as with all such reunions after so many years, our first startling and embarrassed impressions involved the physical changes—the gain in weight, the graying or loss of hair, the double chins. We were kids then and grandparents now. No matter: The old feelings of being caught up together in something so historic and larger than ourselves flooded back, lifting us to that delirious and scary moment when we first acted on our principles and tasted what it was like to take charge of our lives.

What we were feeling was the opposite of nostalgia, which Abbie Hoffman—may he rest in peace—used to describe as a minor form of depression. We introduced ourselves to each other and found that all the people present had gone on, in a dazzling variety of ways, to create lives that more than fulfilled the promise of what we started in those buildings. We had become writers, artists, teachers, community organizers, public-interest attorneys, scientists, judges, scholars, physicians, carpenters, farmers, environmentalists, journalists—all working in our own ways to continue the idealism inherent in our rebellion forty years before. Anyone who's ever said that the generation of the sixties sold out should have attended.

For me the weekend was both a homecoming and a relief, a chance to be among my old comrades and to interact with them as I had before I became "Mark Rudd," the iconic student leader and New Left media star. We talked about our roles in the strike as equals, which is what we were. The balance was righted.

Tom Hayden, after speaking on the opening panel, greeted me by asking, "What's going on here? Have you been purged?" He had noticed I wasn't a featured speaker.

I replied, "The statute of limitations on Mark Rudd has expired."

It was the first time, too, in forty years, in which I felt physically comfortable on the Columbia campus. I had been back several dozen times since 1968—to show my children and my wife and friends from New Mexico around the city, to meet people on campus, to speak at classes—yet I had always experienced the old tension and anger over the way Columbia had treated us and resentment about the enormous power this imperial university wielded in the world. Strangely, this time I felt at home, as you should in a place where you've spent three of the most formative years of your life.

For two days a series of panels made up of scholars—some of whom were participants in 1968, others of whom had not yet been born—commented on topics such as "Columbia 1968 and the World,"

"From Vietnam to Iraq," "Race at Columbia, Then and Now," and "The Legacy of the Student Movement." These sessions became compelling two-hour discussions linking the past with the present, involving whole auditoriums filled with hundreds of people. None of us, it seemed, had lost our taste for talking about politics, society, and culture. On the last night, a dozen well-known writers who as students had participated in the strike read their works, to great applause.

The commemorative aspect of the event, however, was overshadowed by a series of startling revelations by African-Americans and white and black women that raised the weekend to the level of catharsis. Approximately fifty former occupants of Hamilton Hall, members of the Student Afro-American Society, attended. That in itself was a minor miracle, attributable to the work of Thulani Davis, a writer and Buddhist priest and the only black member of the reunion organizing committee. In 1968 she had been a Barnard sophomore in Hamilton Hall. Over the course of months, with a vision that never wavered, she was able to convince reluctant former SAS members that their story would be heard at the event. Many had negative feelings toward Columbia and didn't want to set foot on campus.

Thulani and Tom Hurwitz wrote and directed what turned out to be a three-and-a-half-hour presentation involving more than forty people, entitled "What Happened?" All of us present in the auditorium, speakers and audience, were transfixed listening to our collective history narrated by us, in our own words.

Because of the segregated nature of Columbia University in 1968, most of the whites had been ignorant of the perspective and experiences of the black students. We didn't know them, nor they us. We especially hadn't known the extent of racism black students had to endure on campus—the daily indignity of being stopped by black university security officers who expected only white people at Columbia, the racist jokes and open hostility from white students, the heartbreak of black athletes kept from playing the sports they had trained for. The situation was particularly grinding for the black students, commented Ray Brown Jr., now a well-known attorney in New Jersey, because the discrimination was covert, unacknowledged, and therefore without rules, unlike that in the segregated South.

We whites hadn't known that the Columbia College class of 1968 was the first to admit any appreciable number of black students—twenty people. By the class of 1971, that number was only up to thirty-five—out of a student body of approximately twenty-eight hundred. Similar numbers prevailed at Barnard. In effect, the African-American students at Columbia were the pioneers to integrate the university in a group. Judge Al Dempsey of Atlanta, class of '69 and a veteran of Hamilton, told the stunned audience, "The time I spent here just about destroyed me. . . . The only thing worse was watching my wife die of breast cancer."

For the last forty years, I've thought that the expulsion of SDS and the other white students from Hamilton in the early morning of April 24 was the product of separatist sentiment arising from black-power ideology among the black students. The story told by the black students at the reunion put the whole question in a different light. Speaker after speaker stressed that he or she had arrived at Columbia having grown up in or inspired by the civil-rights movement in both the South and the North. They came to Hamilton Hall with a much stronger sense of the seriousness of the situation and the necessity for discipline. They saw SDS and the other white students in Hamilton Hall during the first hours—with our guitars and our pictures of Lenin—as a bunch of disorganized bozos (my description) without political or tactical coherence (Ray Brown's words). Khalil Rashid of SAS, now a media specialist in New York City, said of black students that they, unlike SDS, "had learned to be measured in confrontation with authority. We also felt that if we suffered because of a confrontation, it would be because of our own actions, and not that of others."

Michelle Patrick, a Barnard student in Hamilton Hall and now a television screenwriter, made a comment that went straight to my heart—and probably that of every other white person present. "The first thing we did, immediately after the whites left Hamilton, was to clean the building from top to bottom. Our parents had taught us never to let whites think of us as dirty. We knew that eventually we'd need to leave the building, and we knew how important it was to leave it *clean*."

She also told us, "We weren't revolting against our parents; we were carrying on the struggle that they had been involved in their whole lives. It appeared as if the white students were in rebellion against their own families."

After the April 30 bust, blacks and whites had gone our separate ways. For many of us, this was the first time in four decades that we had listened to the Hamilton Hall students, and what they told us humbled us in our ignorance. At the end of the presentation, we rushed to embrace each other and to talk more. Everyone present thanked Thulani Davis and congratulated her for her accomplishment.

Listening to the Hamilton Hall veterans that night, it hit me once again, the injustice of Mark Rudd's having become the icon for the strike and SDS's getting the credit and media recognition. In any retelling, the African-American students should have been central. In their place I would have been resentful and angry. The next day I said to Khalil Rashid, "I wish it hadn't taken forty years for us to begin to talk to each other."

He replied, "It took forty years of wandering in the desert to reach the promised land. It's biblical. It took us that long to be able to understand each other."

Other imbalances were righted. All the recognized leaders of the strike in 1968—black and white—were men. At the reunion, women spoke of the fact that their contributions had never been acknowledged and that they were marginalized. They did the lion's share of the grunt work in the buildings—the organizing, the cleaning up, the cooking. Out of 700 arrested in the first bust, 250 were women.

At a panel called "Feminist Legacies of 1968," on Friday afternoon, numerous Barnard women told of the medieval and paternalistic parietal rules that they lived under before the strike. For example, women were daily locked into their dorms at 10:30 P.M. During the strike, Barnard administrators locked the campus gates to keep all the women in. Susan Brown and Barbara Bernstein, both Barnard freshmen at the time, told of climbing out a second-story window and down the side of the building in order to cross Broadway to return to President Kirk's office in Low Library.

At the "What Happened?" presentation, Bonnie Wildorf and Ellen Goldberg, respectively twenty- and eighteen-year-old SDS members from Barnard, divulged for the first time—with great glee—that it was they who had shouted, "To the gym! To the gym!" on April 23, in front of Low Library, when I and the other "leaders" had had no idea what to do. Their disclosure prompted a laughing, cheering ovation.

The sole female member of the Strike Coordinating Committee during the occupation, Rusti Eisenberg of Fayerweather Hall, introduced a sobering note when she told the session, "As a woman, a graduate student, a person new to the Columbia campus, and a spokesperson for the dissenters [on the amnesty demand], I was an unwelcome presence in the Strike Coordination Committee. At the time I was hurt and stunned by the machismo and disrespect of the young men in that group. When I think back, the notable exception, the person that I most remember for his sensitivity and thoughtfulness despite our political differences, was David Gilbert."

Our reunion coincided with five days of student demonstrations against the war in Iraq. Daily, young people took to the Sundial to read aloud the names of the dead, both Americans and Iraqis. They placed mock coffins draped with American flags on the ground in front of them. One day, concurring with a walkout from classes, someone anonymously dropped a huge banner across the front of Butler Library that read, STOP THE WAR, DIVEST. It referred to the students' demand that Columbia sell off all its investments in corporations that produced weapons for the U.S. military. At the same time, protesting students covered the head of Alma Mater, the famous statue on the steps leading up to Low Library, with a hood, recalling the iconic photograph of a torture victim at Abu Ghraib. About fifty people linked arms surrounding the statue to maintain the hood in place for about two hours.

Also that weekend a few dozen students and community residents hooked up once again, as we had forty years before, to try to stop Columbia's current expansion into Harlem, this time northward in a factory and warehouse area known as Manhattanville. They charged that Columbia would advance the gentrification of Harlem, driving out all affordable housing, and also that it was planning to build a hazardous biological lab.

The students who participated in these demonstrations were animated with the same passion that we had forty years before. An anthropology graduate student, Callie Maidhof, told our Vietnam and Iraq panel that antiwar organizing at Columbia was only beginning. Another student, Johanna Ocaña, whose parents were immigrants, spoke on the "Race at Columbia" panel about ongoing support for immigrant rights.

On Sunday morning, the last day of the event, we held a memorial at Columbia's religious center, Earl Hall, to remember all those who had passed on in the intervening forty years. We read a list of approximately seventy-five names—numerous comrades from the strike, some of our professors, "outside agitators" like Abbie Hoffman and Stokely Carmichael, and even David Truman, our administration nemesis. We also included a few people who had inspired us, like Malcolm X.

People memorialized our friends who had passed. JJ was hailed as an anti-imperialist fighter whose ashes were spread at the tomb of Che Guevara in Cuba; Ted Gold was honored as a great organizer with a fabulous outside jump shot. My own contribution was to join my old friend Jeff Sokolow in reciting the Kaddish, the Jewish prayer for the dead. It's the only thing I know how to do when someone dies. The memorial was opened and closed by James "Plunky" Branch, a professional jazz musician who had been in Hamilton Hall, playing soaring tenor-sax improvisations of "Amazing Grace" and "Take My Hand, Precious Lord."

On Sunday afternoon a few dozen people from the reunion gathered at the former gym site in Morningside Park. Forty years ago the park was an abandoned no-man's-land, with two and a half acres scarred by heavy construction equipment. Now, on this rainy spring afternoon, flowers and trees blossomed in a restored treasure of a park originally designed in the late nineteenth century by Frederick Law Olmsted, the master designer of Central Park.

The organized communities of Morningside Heights and West Harlem and the students had stopped Columbia from building a segregated gym on public land. Sam White, a Harlem resident then and now, who was an SAS member in Hamilton Hall as a young man, praised the "creative obstruction" of the takeover and the support of the Harlem community people who had come up to Hamilton to provide food and protection. The final speaker was Suki Ports, a Morningside Heights resident who, forty years before, had sat in the scoop of a giant bulldozer to keep it from knocking down trees "that it took three of us to put our arms around." Everyone present celebrated the fact that the precious Manhattan park had been saved from a ten-story concrete monstrosity.

Our reunion's gift to the park and the community was a weeping cherry tree, which Ms. Ports and Sam White helped plant next to a waterfall running down a rock to a large landscaped pond below. Forty years before, the pond had been a gaping hole excavated for the gym's foundation.

Acknowledgments

My sincere thanks to the many friends who have helped and encouraged me over the years it took to write this book. I apologize for not naming you individually. Alas, I have not kept any accurate record as to who read which of the various drafts, who suggested which changes, who bolstered me at low times with the thought that the project was worthwhile. Not wanting to leave anyone out, I can best thank you collectively for your time and energy and ideas.

Similarly, I can't possibly name all the people who took the risk of aiding and harboring me during my time as a fugitive. You know who you are.

I would like to remember my old comrades who didn't make it out of early adulthood—Diana Oughton, Ted Gold, and Terry Robbins; similarly, David Gilbert and Judy Clark, who remain in prison.

If not for the five years of their lives that Sam Green and

Bill Siegel donated to making *The Weather Underground* documentary, I would never have told this story.

Special thanks to my literary agent, Jane Dystel, who over more than two decades never lost faith in the eventual existence of this book, and to my editors at William Morrow, Henry Ferris, his assistant, Peter Hubbard, and master copyeditor Maureen Sugden, who all helped turn a mountain of a manuscript into a work I'm proud of.

Underground is dedicated to my parents—Jake, who passed away in 1995, and Bertha, still very much alive and as sharp as ever at age ninety-seven. Their unconditional love kept me going. In this book I hope to answer my mother's eternal question, "How could you do this to me?" This book is also for my children, Paul and Elena, who have long wondered, "What was all that about?" And for my brother, David, who bore the brunt of my parents' anguish for seven years and enthusiastically encouraged me to write this book.

Thanks to my dear wife and comrade, Marla Painter, for the help and love that sustained me during the long writing.

Finally, *Underground* is dedicated to my ex-wife and partner for the seven years underground, Sue LeGrand. Without her I would not be alive today.